Animal Testing and Consumer Products

HEIDI J. WELSH

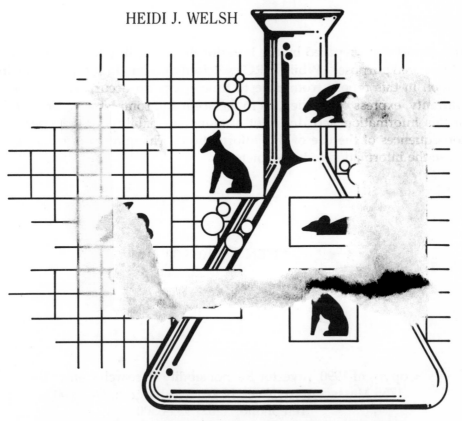

INVESTOR RESPONSIBILITY RESEARCH CENTER

This report was written by Heidi J. Welsh. Charine Adams prepared it for publication, and Michael J. Davis designed the cover.

The Investor Responsibility Research Center compiles and impartially analyzes information on the activities of business in society, on the activities of institutional investors, on efforts to influence such activities, and on related public policies. IRRC's publications and other services are available by subscription or individually. IRRC's work is financed primarily by annual subscription fees paid by some 400 investing institutions for the Social Issues Service, the Corporate Governance Service and the South Africa Review Service. This report is a publication of the Social Issues Service. The Center was founded in 1972 as an independent, not-for-profit corporation. It is governed by a 21-member board of directors who represent subscribing institutions.

Executive Director: Margaret Carroll
Social Issues Service Director: Carolyn Mathiasen

ISBN 0-931035-39

Table of Contents

INTRODUCTION

The burgeoning animal protection movement has turned a spotlight on major U.S. corporations and their use of animals to test consumer products. Activists remain critical of corporations' progress and commitment to stop animal testing, although in the last decade firms have spent millions on research on alternative tests and now use fewer animals. The strident debate has traveled from sidewalk demonstrations to university classrooms, and in 1987 it gained official entrance to the annual meetings of three companies in shareholder resolutions from animal rights activists. Since then, research has intensified, boycotts and public relations campaigns have been waged, more shareholder proposals have surfaced and several major cosmetics companies have stopped animal testing.

In the mid-1970s, the concept that animals have rights inspired a new breed of political activist to launch a moral assault on society's ideas about the appropriate tenor of human and animal relations. Now, a decade and a half later, the new movement has consolidated remarkable support from a wide range of sources. Government regulatory agencies have been scrutinized for their ambiguous role in requiring animal tests for consumer product safety, companies have been forced to justify their use of animals, and scientists have created a whole new field to assess product safety without using animals.

Showing how mainstream concern on this issue has become, in a *Newsweek* column in April 1989, *Washington Post* editorial page editor Meg Greenfield wrote, "We have become far too self-indulgent, hardened, careless and cruel in the pain we routinely inflict upon animals for the most frivolous, unworthy purposes." Greenfield did not come out in favor of abolishing animal research, but she questioned the merit of testing for new consumer products to add "to our already obscene store of luxuries and utterly superfluous vanity items." Prominent animal rights activist Henry Spira concluded to IRRC that there has been a "remarkable change in our culture that says animal suffering does matter. People relate to animals in a different way now."

This book looks at the animal protection movement and its impact on the consumer product companies that have been its primary targets. For the purposes of this report, consumer product testing includes the toxicity testing of cosmetics and household products that are not medical-related. It is important to keep in mind that such testing accounts for only a small portion of the estimated 20-70 million laboratory animals used every year in the United States. Most experimental animal use is for biomedical research or for the development and testing of new pharmaceuticals. This book classifies the testing of pharmaceuticals and over-the-counter drugs as medical-related testing and does not directly cover that topic. Further, the report does not specifically discuss basic biomedical research conducted by scientists at private and public facilities. (The analysis of animal use patterns in Chapter V does cover all types of animal use, however.) Despite the limited focus, many of the issues raised in this book are applicable to the more controversial debate over a ban on animal use in medical-related testing and research, which many in the corporate and scientific community fear is the ultimate goal of the activists who are now limiting their fight to consumer product testing.

Overview of Chapters

Chapter I looks at the evolution of the animal protection movement, contrasts newer animal rights philosophy to more traditional animal welfare ideas, and examines the reaction by companies and proponents of animal research to the movement. This chapter also describes activists' use of the shareholder resolution tactic.

Chapter II examines the array of federal laws and regulations that govern animal testing. An analysis of federal and state legislative battles shows where the main actors in the debate stand.

Chapter III covers animal tests and their alternatives. There is a description of the kinds of techniques now in use for product testing and an examination of what sorts of alternatives now are used or being developed. The validation of these nonanimal tests is discussed in the conclusion of this chapter.

In-depth profiles of nine U.S. companies that have received shareholder resolutions on animal testing make up Chapter IV. The profiles provide data on the corporations' animal use and discuss their work on finding and implementing alternatives.

Chapter V is an analysis of data submitted to the U.S. Department of Agriculture by all research facilities that reported animal use in fiscal years 1986-1988; IRRC obtained the data using Freedom of Information Act requests. Tables show animal use by species and experiment type, for commercial and non-commercial users. This chapter also looks briefly at federal reporting requirements for painful experiments.

Included as an Appendix is a directory of 47 nationally focused animal protection groups that responded to a questionnaire from IRRC. Other Appendices provide a listing of scientific groups that are coordinating the search for alternatives to animal tests and the industry associations that have concerned themselves with this issue.

* * * *

This report was written by Heidi J. Welsh, a research analyst at IRRC. She would like to thank Dr. Alan Goldberg of the Johns Hopkins Center for Alternatives to Animal Testing and Drs. Andrew Rowan and Karl Andrutis of the Tufts University Center for Animals and Public Policy, who took time out of their busy schedules to read and comment on Chapter III. She also thanks Carolyn Mathiasen for her cheerful encouragement and editing, Duane Walker for his patience and unflappable ability to answer computer questions, Charine Adams for her perseverence and skill in preparing the manuscript, Shirley Carpenter for her Harvard Graphics expertise and Geraldine Rowe and Linda George of PSSG for data entry.

I. THE ANIMAL PROTECTION MOVEMENT

The notable increase in attention to the use of animals in product safety testing is tied directly to the remarkable growth in the animal protection movement in the last decade. This chapter describes the movement today, provides a synopsis of the recent history of animal protection activism and gives an overview of animal rights philosophy, with an emphasis on events related to animal testing.

The animal protection movement encompasses organizations as disparate as the Animal Liberation Front, designated a terrorist group by the FBI, and the American Society for the Prevention of Cruelty to Animals, a mainstream animal welfare charity. While the ALF undertakes illegal laboratory break-ins to "rescue" research animals, the ASPCA employs considerably more moderate tactics such as newspaper ads to encourage better animal treatment. In between these groups is the rest of the movement whose various members have broadcast concern for veal calves, vilified fur retailers, castigated medical researchers and protested hunting.

Taken together, humane or welfare groups combined with rights organizations form what is commonly called the animal protection movement. All observers agree that membership in animal protection groups has skyrocketed in the last decade, but estimates vary widely: Activists claim 10 million adherents nationwide, while industry and biomedical research support associations give a much lower number. The *Animal's Voice* magazine, one of at least six periodicals for the movement, lists in its February 1989 issue 104 groups that are "the animals' allies" around the United States, as well as 14 groups in foreign countries.

In an April 1989 article in the *Chronicle of Higher Education,* Dr. Andrew Rowan, head of the Center for Animals and Public Policy at Tufts University, reported that membership in animal rights groups expanded five- to ten-fold in the 1980s. One notable case is People for the Ethical Treatment of Animals, now the largest radical rights organization in the United States; PETA started in 1980 with 100 members and now says it has close to 300,000 in its ranks. In a cover story on the animal protection movement at the end of 1988, *Newsweek* estimated that there were some 7,000 groups in the United States, with combined budgets totaling about $50 million. The financial resources of such groups are probably much greater, however; 29 of 47 groups surveyed by IRRC that responded to inquiries about their finances reported about $49 million in combined annual budgets.

The *Animals' Agenda,* a national animal rights magazine, gave potential advertisers a self-described snapshot of its readers in its January/February 1990 issue that probably serves as a good description of the rights component of the movement. The magazine says it has an audience of about 75,000; 80 percent are in professional or business-related occupations, 75 percent are women, 84 percent are college graduates and 25 percent have M.A. or Ph.D. degrees.

The Evolution of the Animal Rights Movement in the United States

Scientists using animals for their research had to fend off virulent antivivisectionist critics in the last few decades of the 19th century in England; at about the same time, antivivisectionism spilled over into the United States. Interest in the cause waned as the century turned, though, and most U.S. citizens who continued to express concern for animals did so by working in the humane movement. By 1966, welfare groups managed to push the Animal Welfare Act through Congress, requiring regulation of research laboratories using certain species and some animal dealers. Subsequent amendments to the act have strengthened it, but animal protectionists have never been satisfied with the law's ability to ensure proper animal care.

In the late 1960s, before animal rights developed its mass appeal, interest in alternatives to animal testing surfaced in both England and the United States. The British Parliament charged a committee to report on the status of alternatives science in 1965, and in 1969 the Fund for the Replacement of Animals in Medical Experiments formed in the United Kingdom to encourage scientific research on alternatives. In this country, activists established United Action for Animals in New York in 1967 to promote alternatives, particularly replacement techniques.

Still, the renewed interest in animals did not shift significantly away from traditional humane society concerns until Australian philosopher Peter Singer published *Animal Liberation* in 1975. Today, Singer's book is widely regarded as the impetus for the current animal rights movement; it provided a coherent philosophy for activists to adopt. Groups began to spring up around the country to combat "speciesism"—what Singer described as "a prejudice or attitude of bias toward the interests of members of one's own species and against those of members of other species."

Sporadic demonstrations began to surface in the United States as the 1970s progressed. Rights activists have chosen between two basic strategies over the years—a relatively moderate, behind-the-scenes effort to initiate change, and dramatic confrontation with all-or-nothing demands. The projects of Henry Spira, one of the chief organizers in the movement, illustrate the incremental approach; although a clear proponent of animal rights, Spira has garnered support from several of the major welfare organizations. PETA's strategy, on the other hand, has been to work outside the fold of welfare groups, depending on sometimes outrageous tactics to seize the public's attention and attract new supporters.

Spira, a former high school teacher, merchant seaman and civil rights activist, took a class about animal rights taught by Singer in New York in 1973 and pulled together class members later to work on the issue. After thorough research, the group settled on protesting sex experiments on cats at the American Museum of Natural History. By 1976, public outrage over the cats' plight forced the museum to stop its experiments. Spira went on to work with other groups in the state to push through a bill in the New York legislature banning the use of dogs and cats from pounds for research. Then, in 1979, turning his sights on the use of rabbits in Draize eye irritancy tests to determine the safety of cosmetics, he organized the Coalition to Abolish the Draize and picked Revlon as its first target. After private efforts to persuade the company to support research on alternatives failed, the coalition persuaded the Millenium Guild, an animal welfare philanthropy in New York, to run full page ads in the *The New York Times*, asking, "How Many Rabbits Does Revlon Blind for Beauty's Sake?" By the end of 1980, the company capitulated; it donated $750,000 to research into Draize alternatives at Rockefeller University and presented a major victory to the new rights movement.

PETA got its start several years after Spira began his work in New York. Alex Pacheco, a student at George Washington University, became interested in animal rights and in 1980 volunteered at a Washington, D.C., animal shelter directed by Ingrid Newkirk. Newkirk read a copy of *Animal Liberation* Pacheco gave her, and the two decided to form PETA to educate the public about animal abuse. They began by promoting vegetarianism and "cruelty free" products not tested on animals. Then, in the summer of 1981, Pacheco went to work as an intern for Dr. Edward Taub, a researcher in the Washington, D.C., suburb of Silver Spring. Taub had a grant from the National Institutes of Health to study neurological trauma in monkeys, which he induced by cutting nerves in various parts of their bodies. Pacheco's documentation of the animals' filthy condition induced local police to seize the monkeys, and information PETA provided to the press brought national coverage to the incident. Taub lost his funding and was convicted on animal cruelty charges that were later overturned on a technicality. With the "Silver Spring monkeys" case, PETA had successfully launched itself into the public eye, where it would continue to grab headlines as one of the loudest voices in a rapidly expanding corps of activists.

In 1981, after continued pressure by animal protection groups, the Cosmetic, Toiletry and Fragrance Association also agreed to fund research on alternatives. The CTFA established the Johns Hopkins Center for Alternatives to Animal Testing with a $1 million grant. Since its founding, CAAT, under the direction of Dr. Alan Goldberg, has come to be recognized by many in the scientific community as the primary clearinghouse in the United States for information on alternative toxicology methods. More radical rights groups discount the Center's efforts as industry-supported tokenism—charges that CAAT and more moderate activists vigorously refute. Whether new toxicology methods are sufficiently advanced to replace all animal tests has remained controversial. Activists insist the technology is available now to stop using animals, while companies and many scientists say that more research is needed before all reliance on animals can be ended, although use of alternatives can reduce this reliance significantly. (Chapter III discusses alternatives more fully.)

At the end of 1983, Henry Spira described the activities of his coalitions, noting continued success with the campaign to find alternatives to the Draize test, and the "immediate impact" of his newer Coalition to Abolish the LD50. (Along with the Draize, the LD50 has been the most controversial animal test.) He noted a growing consensus in regulatory and industry circles that the classical LD50, in which half the test population of about 100 animals die, was outdated. "We forge ahead with a flexible, step-by-step movement towards animal rights," he wrote. Spira concluded that his incremental approach seemed to be having more effect than other tactics: "The 100 years of hysterical and self-righteous demands for immediate abolition [of animal use] has led to neither short-term nor long-term results."

Despite Spira's assessment, PETA continued its work in the same vein it had already established. In summer 1985, it used videos obtained by activists in an illegal raid to document animal abuse at the University of Pennsylvania, again making the nation's headlines. The public outcry that resulted from the scenes showing lab workers' interaction with baboons used in violent head injury experiments caused the National Institutes of Health to cut funding for the project.

Also in 1985, PETA began to take on consumer product testing. Ingrid Newkirk told IRRC this strategy was formulated after she was taken to a testing laboratory in New Jersey "quietly by someone who worked there." Newkirk said she was appalled at the number of animals she saw and wanted to do something about it. With its new "Compassion Campaign," PETA distributed literature and asked people to use the companies' refund policies to return products tested on animals. Activists also picketed the annual meetings of Noxell, Avon Products and International Playtex. Newkirk said the new campaign was not a retreat from PETA's previous work against animal use in medical research, but that "our work with consumers is most important because consumers dictate what will happen to animals" by their buying habits. She asserted that if PETA can produce an ethical shift in the consuming public, companies will have to change.

PETA's concern with consumer product testing eventually led to its use of shareholder resolutions to gain official entrance to the annual meetings of nine major companies to protest their animal use, starting in 1987. Since then, other animal protection activists have used shareholder resolutions to address factory farming, the alleged unnecessary use of animals in medical product testing and the selling of fur coats, in addition to animal testing. (The section starting on p. 20 describes activists' use of shareholder resolutions more fully.)

As 1986 ended, Spira again took stock of animal protection efforts. He mentioned a survey from the Food and Drug Administration showing a 96 percent decrease in the use of the classic LD50 and noted that the Soap and Detergent Association and Bausch & Lomb were funding research on Draize alternatives. Key to this progress, he felt, was the strategy to create "a cascade effect by planning a sequence of ever-enlarging winnable battles which were often based on ideas suggested by the science community itself, thereby gaining not only the support of the general public, but of scientists as well." He wrote, "We always began with discussion" but "did not hesitate to play hard ball when warranted."

One of the main controversies that helped to provoke public debate over animal tests for consumer products occurred at Gillette. During the fall of 1986 the now-defunct radical animal rights group Ark II called a boycott of Gillette to force it to stop testing products on animals. Ark II contended that animals in Gillette's custody were abused, backing up its allegations with information gathered by Leslie Fain, a technician who had worked in Gillette's laboratory. Gillette denied Fain's charges and disputed her credibility. The company now says PETA may have placed Fain at the lab to generate publicity for its Compassion Campaign. At the time of the Ark II case, the company told IRRC that "statements and video clips in recent news releases are grossly misleading," and that some of the videos shown contained scenes that did not occur at Gillette. In February 1987, an Agriculture Department investigator concluded that although some handling violations occurred, the abuse described by Ark II was not apparent. Eventually, however, the USDA sent a warning letter to Gillette based on Fain's evidence. Gillette closed its controversial animal research facility in May 1987; the company said extensive corporate restructuring forced by a takeover attempt from Revlon was the reason for the laboratory's closure, but activists claimed victory.

In 1988, media attention focused on two highly visible cases that illustrated the increasing intensity evident on both sides of the controversy over animal use in general. In September, Cornell University researcher Michiko Okamoto returned a $530,000 grant from the National Institute on Drug Abuse for the continuation of her 14 year study into barbiturate addiction in which she used cats. The New York rights group Trans-Species Unlimited had targeted Okamoto for months with pickets, letters and personal contacts. Two months after Okamoto conceded, a New York woman, Fran Stephanie Trutt, was charged in a bombing attempt at the Connecticut facility of a small public company, U.S. Surgical Corp., which uses dogs to demonstrate its surgical staplers. Showing how far corporate concern can go, the company's president, Leon Hirsch, revealed in January 1989 that U.S. Surgical had paid informants in animal rights groups since the early 1980s; animal rights groups since have alleged Trutt was entrapped by one of the company's informants. (See box.)

Among its many other projects, PETA has continued its focus on consumer product testing with its Compassion Campaign, concentrating on several major companies. A 1988 boycott of Benetton that had extended to Europe ended in November of that year when the company said it would stop animal testing. In June 1989, a four-month boycott against Avon concluded when the company announced it had stopped its use of animals, and an infant campaign against Revlon died when that company ended animal testing as well. By December, however, PETA was considering a resumption of its earlier effort against Avon, because the company said it might use materials from chemical suppliers that test on animals, although it encouraged the suppliers not to use animal tests. Boycotts of Gillette and Cosmair (L'Oreal) announced in the summer of 1989 continued as 1990 began, and early in 1990 Susan Rich of PETA went to France to organize a European flank to the attack on Cosmair.

Led by Spira, the less radical wing of the animal protection community has expressed disapproval of PETA's choice of Avon and the fact that it may continue to approach the company over the supplier issue. Given the company's record since 1981 of funding extensive research on alternatives, Spira says Avon is just not an appropriate target. The focus on Avon "is undermining the alternatives effort. One major company has said it was thinking of getting out of alternatives work because of PETA's tactics," Spira told IRRC. Companies that get a high profile from their alternatives work are being attacked by PETA and others to do still more, and they do not want the adverse publicity that results from an Avon-like boycott—they think they should get credit for what they have done, Spira said. In December 1989, the ASPCA and the Humane Society of the United States joined with Spira in a resolution that said, "We believe that it is counterproductive to engage in boycotts or other negative actions against responsive companies and that such tactics [should] be used only against corporations or institutions which refuse to deal in a constructive manner with the animal protection movement."

PETA maintains that its tactics are working, so there is no reason to alter them. It says it does not support the incrementalist tack because "the animals don't have time to wait." Newkirk told IRRC that although she admires Spira for his initial work at publicizing the issue, she feels he has become too "friendly with the companies" and should step aside. In a less critical voice, Kathy Guillermo, director of PETA's Compassion Campaign, said she saw all of the different tactics of the movement as playing a role in bringing an end to animal testing.

Whatever the internal arguments of the movement, it is clear that animal rights activists have brought a higher profile and greater sense of urgency to animal issues in the last decade, and have had a clear impact on the use of animals in research and testing. While PETA's very public attacks have aroused strong public concern that has forced companies to address the issue, these actions also have alienated firms and may have undermined PETA's ultimate effectiveness. Spira has been extremely effective as well, by engaging in dialogue with the companies, but he has been criticized for selling out to an extent, because of these tactics and his unwillingness to target responsive corporations. Both strategies of the movement continue to present a united front in criticizing unnecessary animal use for testing consumer products, however. As the 1990s begin, the movement's continued clout has caused companies and supporters of biomedical research to issue dire warnings about its impact on their work.

Reaction by Industry and Animal Use Proponents

The opponents of the animal rights movement seem particularly concerned

Activists allege setup

Fran Stephanie Trutt was arrested for attempting to plant a bomb at U.S. Surgical Corp. headquarters in Norwalk, Conn., early on Nov. 11, 1988. Trutt was unable to meet bail set at $500,000 and was held on charges of attempted murder (of Leon Hirsch, the company's president), possession of explosives and manufacture of a bomb. Although most animal protection activists initially kept their distance from Trutt, they later said that U.S. Surgical set her up in an attempt to discredit protesters who had been picketing the company for seven years.

The facts of the case remained muddied, with bitter and contradictory accusations from both sides. In January 1989, information leaked out that a woman connected with a security firm hired by U.S. Surgical, Perceptions International, had befriended Trutt and tape recorded conversations in which she threatened to bomb the company, and that another Perceptions International employee had driven Trutt and the bomb to U.S. Surgical after notifying the police. In a March 3, 1989, article, *The New York Times* reported that U.S. Surgical president Hirsch defended the undercover operation. Hirsch was quoted as saying, "Don't forget: This lady tried to premeditatively murder me." He told the *Times*, "We have felt since the early 1980s that the animal rights movement was edging toward violence, and to protect our property and our people, we started developing surveillance systems." Animal rights activists, though, said that U.S. Surgical crossed the line into entrapment.

The head of Perceptions, Jan Reber, was arrested in October 1989 on felony charges of conducting an illegal investigation and operating a private detective agency without a license. Further, the Perceptions employee who drove Trutt to the company with the bomb, Marc Mead, is in federal prison in Pennsylvania. When Mead picked up Trutt in New York City,

about activists' work on proposed legislation that would ban either the Draize or the LD50. At a press conference in the fall of 1989, Edward Kavanaugh, president of the Cosmetic, Toiletry and Fragrance Association, said, "American consumers are under attack; the scientific process which protects consumers is the real victim of this unnecessary and dangerous campaign." He asserted that although his organization would continue to support research into alternatives to animal testing, technology needed to replace all animal use was

in U.S. Surgical bombing

he violated his parole from an earlier offense by leaving Connecticut.

As of January 1990, Trutt was still in prison, at the Niantic Correctional Facility in Connecticut. A judge ruling on the federal charges--possession of explosives--had sentenced her to three years of probation and ordered her to undergo psychiatric counseling. The state trial on the alleged attempted murder was scheduled for the end of February 1990.

Friends of Animals, the Norwalk, Conn., organization that has led protests against U.S. Surgical's use of dogs since 1982, alleged that U.S. Surgical manipulated Trutt into attempted violence. At Trutt's federal sentencing, the state prosecutor submitted a brief that questioned the circumstances surrounding her arrest, a representative of Friends of Animals told IRRC.

At the same time, the group said Trutt was not a member, and it dissociated itself from the use of violence. In its April/May 1989 newsletter, the group concluded: "For seven years we have actively, peacefully protested this company's use of living dogs in surgical demonstrations of its products for sales purposes."

In the furor over the Trutt case, activists wondered what other measures companies might be using to counteract them. Mary Lou Sapone, the Perceptions International employee who allegedly befriended Trutt and then tape recorded her conversations, was said to have shown up at animal protection rallies around the country. An article in *The Advocate*, a Stamford, Conn., paper, stated: "Activists said they have evidence that Mary Lou Sapone infiltrated their ranks as a paid informant for the biomedical research community and an investigative arm of the federal government. They say her information gathering, and subsequent leaks, have extended far beyond the interests of U.S. Surgical."

not yet available. Animal tests therefore must be kept open as an option, the CTFA says. Kavanaugh contended that proposed legislation "would knock down what is essentially the first domino of animal testing benefit, behind which are the dominoes of medical and biological research, as well as other consumer product development and safety research." (See Chapter II for more on the proposed legislation.)

The CTFA's opposition to legislation governing consumer product tests is shared by a number of other organizations. In 1979, to counteract what they saw as threatening events, proponents of animal use formed a lobbying group, the National Association for Biomedical Research, and a nonprofit sister organization, the Foundation for Biomedical Research. Since then, the NABR/FBR has worked to present an alternative view to animal rights thought, producing literature and testifying vehemently against bills at the state and national level that would restrict product testing. Barbara Rich of the NABR told IRRC that her association has about 320 institutional members, some 70 percent of which are universities with medical schools or substantial biomedical research programs, and 30 percent of which are corporations, mostly pharmaceutical companies. Their common interest, Rich said, is the "appropriate use of animals." Although NABR's work mostly concerns animals in research, she said there is some focus on consumer product testing as well.

In a manual designed to counter "animal research crises" that the NABR circulated to its members in January 1989, the association stated:

> The animal rights movement has had an enormous impact on the research community...in the United States in the last decade....By combining visceral images with spoken half-truths, lies and emotional invective, they have attempted to create an image of biomedical researchers as uncaring, profit- and/or grant-motivated perpetrators of unspeakable atrocities on innocent animals, especially pets, all for the sake of useless research.

The NABR says "the scientific community did not initially recognize the significant shift in the nature of animal interest groups" that occurred 10 years ago. Rich feels animal protectionists' focus on consumer products is only the first item on their agenda, and that a ban on medical research that uses animals is clearly their ultimate goal. The growing influence of animal rights activists has pushed the whole animal protection community to adopt "a more confrontational strategy," the NABR wrote.

To counter this confrontation, the opponents of animal activists are girding themselves for a long battle. IRRC obtained a copy of an "action plan" to neutralize the impact of animal rights influence that the American Medical Association circulated in June 1989. The version IRRC received came from the Physicians Committee for Responsible Medicine, a group of physicians that promotes alternatives to the use of animals in research and testing. IRRC requested a copy of the document from the AMA, but received a different, relatively neutral assessment of the movement and the scientific community's response—perhaps the "white paper" that the purported "action plan" said should be distributed as part of an extensive public relations campaign by the AMA.

The action plan says "hardcore activists" are at the center of the movement, but that their extreme goals are "not necessarily fully shared by those less fully committed." Therefore, "To defeat the animal rights movement, one has to peel away the outermost layers of support and isolate the hardcore activists and shrink the size of the sympathizers." This can be done by exploiting the differences within the movement, the document says, by attacking "the violent elements." Also, it says, the medical research community must show that it supports humane treatment of its animals, to neutralize this area as a vulnerable point. Finally, the document says researchers must stop being reticent about addressing the issue for fear of being targeted by the activists, because "this strategy is not working and must be changed. The frontal attack on biomedical research can be won only through a strong concerted effort...."

The AMA's action plan says a public relations firm should be consulted for a broad-based education effort, with help from members of the health care community, that would be aimed at the general public but pay special attention to children and teachers. Further, the plan proposes that legal strategies should be developed for "preventing restrictive legislation and protecting freedom of scientific inquiry." Suggested goals include contesting animal rights groups' tax-exempt status, promoting federal investigation of animal rights activities, and lobbying for the "creation of a Justice Department database to monitor and prosecute illegal activities of animal rights groups," as well as compiling a "private database of animal rights activities."

The NABR says information already provided to the public by scientists and patients about the human applications of animal research and researchers' efforts to minimize pain in research animals has blunted the influence of animal rights activists. It says activists recently have backed away somewhat from attacks on medical research, and instead have "honed in on the less sympathetic uses of animals: consumer product testing, basic research with no clear clinical benefit, behavioral research, pound seizure and the use of animals in secondary school education." Despite the apparent popularity of animal rights, the NABR contends that "the overwhelming majority of Americans" express "basic support...for the use of animals in biomedical research and testing, whether disease-related or for product safety."

Animal Rights Philosophy

Animal rights philosophy presents a radical reinterpretation of the human relationship to animals that is distinct from traditional animal welfare views and a departure from common Western beliefs on this subject.

Animal welfare proponents say simply that people should not treat animals cruelly, but they do not call for the total abolition of all animal use by humans.

The credo of the Humane Society of the United States states that "Every living creature has an intrinsic value that derives from creation." Since humans are "uniquely endowed with a sense of moral values," the society says, they are obligated to look after the welfare of domesticated animals and "those upon whose natural environment [they] encroach." The ASPCA, with a similar philosophy, concerns itself with pets, factory farming, zoos, and animals in entertainment, work and education. Both organizations oppose the use of pound animals for research and feel that animal research should be allowed only if all efforts are made to eliminate pain and suffering and alternative methods are not available.

While animal rights activists support most of the work done by humane groups, their adherence to an intrinsically more radical viewpoint sets them somewhat apart from these traditional welfare organizations. PETA, for instance, does not support efforts to change the living conditions of veal calves, because it opposes eating animals in the first place.

Animal rights thinking challenges current Western views of animal and human interaction that have their roots in the Bible. The book of Genesis describes the creation of human beings and the granting of their dominion over all the other creatures of the earth. According to the Judeo-Christian view, this relationship implies responsibilities. "We have the option to decide to dominate animals, but we also have a mandate to make choices responsible to comply with the obligations of stewardship," the National Research Council of the National Academy of Sciences concluded in a 1988 report on animal use in research. In the animal rights view, however, stewardship is another term for "speciesist" oppression.

The ideas of 17th century French philosopher Rene Descartes marked the "absolute nadir" of speciesism, according to the rights movement's Peter Singer. Descartes saw animals as automata, without mind or soul, and said their reactions to painful stimuli were only mechanical. He said humans are distinguished from animals by their ability to reason and to experience emotion and suffering. English philosopher Jeremy Bentham (1748-1832), one of the primary ideological architects of Victorian antivivisectionist thought, rejected Descartes's thesis. Bentham held that living creatures can suffer and enjoy, and that their inability to reason in the Cartesian sense is irrelevant to the moral issue of how they should be treated. Bentham's oft-quoted summation of his philosophy on animals, "The question is not, can they **reason**? nor, can they **talk**? but, can they **suffer**?" appeared on the banners of animal advocates in the late nineteenth century and survived the leap to today's animal rights posters.

In establishing the ethical basis for the current concept of animal rights in

Animal Liberation (1975), Singer elaborated on Bentham's ideas. He said, "I have never made the absurd claim that there are no significant differences between normal adult humans and other animals. My point is not that animals are capable of acting morally, but that the moral principle of equal consideration of interests applies to them as it applies to humans." Singer concluded that the "speciesism" predominant in society today must be ended:

> Will our tyranny continue, proving that we really are the selfish tyrants that the most cynical of poets and philosophers have always said we are? Or will we rise to the challenge and prove our capacity for genuine altruism, by ending our ruthless exploitation of the species in our power, not because we are forced to do so by rebels or terrorists, but because we recognize that our position is morally indefensible?

Tom Regan, a philosophy professor at North Carolina State University, has been termed "the guru of animal rights" by *Esquire* magazine. Regan sets up detailed arguments for a basic animal rights philosophy in his 400 page book, *The Case for Animal Rights* (1983). He extends Singer's argument, saying that since all animals, regardless of species, have inherent value, this means "all have an equal right to be treated in ways that do not reduce them to the status of things, as if they existed as resources for others." Regan presents a somewhat different view of animals than does Singer, by viewing them as "moral patients," equivalent to babies and mentally handicapped humans. In spite of animals' intellectual inferiority, Regan says, we do not have the right to use them as we please, any more than we ought to experiment on the mentally handicapped.

At the end of his book, Regan outlines the implications the rights view has for product testing, first dividing products into nontherapeutic (household products that are not medical-related) and therapeutic (medical-related) categories. His use of "product" as described here concerns nontherapeutic items only. Product testing cannot be justified by saying that it will ensure safety, he argues, because this assumes new products will be manufactured, whereas there is no compelling reason for new products to be made, because we have enough already. "The central moral point, then, is this: No consumer will be made worse off than any test animal if no new products are introduced," he says. Regan maintains that using federal requirements as a justification for tests does not work for the same reason, since although government agencies require or encourage tests for new products, they do not require the creation of new products. Furthermore, if company profits fall because firms cannot test products on animals, the moral equation is not affected, Regan argues. The harm caused to company officials is not comparable to that caused to the test animals, and besides, "Those who voluntarily participate in a business venture voluntarily take certain risks, including the risk that they might be made worse-off if their business fails."

Product tests are wrong "because they violate the rights of laboratory animals;...the basic moral right of these animals not to be harmed," Regan asserts. The options left to corporations, in his view, are to compete with existing products and to "outdo one another by developing nonanimal alternatives." These conditions would create competition that "would be a paradigm of the free enterprise system at its finest," he concludes.

In its 1988 report on animals used in research, the National Research Council presented a view of animal and human interaction that contrasts sharply with Regan's. The council said the concept of inherent rights has never been applied to nonhumans before. Although there is common recognition of the "inherent value" of animals and a consequent obligation not to cause them unnecessary pain, the council found, "From tradition and practice it is clear that society accepts the idea of a hierarchy of species in its attitude toward and its regulation of the relationships between humans and the other animal species....Clearly, humans are different."

Marjorie Spiegel counters this view in her book, *The Dreaded Comparison, Human and Animal Slavery* (1988), by equating speciesism with racism. Using a series of historical quotes and disturbing pictures comparing the mistreatment of blacks and animals, Spiegel builds her case. She notes that blacks were viewed as animals by slaveholders: "The ethic of human domination removed animals from the sphere of human concern. But it also legitimized the ill-treatment of humans who were in a supposedly animal condition." Alice Walker, author of *The Color Purple*, sums up Spiegel's thesis in her preface to the book, injecting a feminist component as well: "The animals of the world exist for their own reasons. They were not made for humans any more than black people were made for white or women for men."

Feminists for Animal Rights, based in Berkeley, also sees a common oppression in the patriarchal domination of the environment, women and animals. "Historically, women have always been viewed as being closer to nature than men. While men were transcending their dependence upon the natural world and building 'sophisticated' cultures, they were simultaneously oppressing women," FAR says.

Animal rights activists clearly draw on environmentalist ethics too. Nevertheless, although some environmentalists have broadened their concerns for wildlife habitats and the health of the biosphere to include laboratory and farm animals, extensive support for animal rights from the environmental community is not yet evident. Both Newkirk and Spira see current "animal-based agriculture" techniques as destructive to the earth. They see support for animal rights as a natural next step for groups such as the European environmental movement, the Greens. Spira feels we need to "reassess our relationship with the natural world"; living in harmony with nature should include respect for animals, and this means "they are not edibles," he says.

A growing emphasis on work against factory farming, using animal rights philosophy merged with an environmentalist justification, is now emerging in the movement; in fact, Spira maintains that rights work will be sustained in the long term by just such a focus. Spira has begun work to persuade a number of other animal protection groups to put their energy primarily into fighting factory farming, since billions of animals are used in agriculture for food—a figure that dwarfs the number of animals used in research and testing.

In *Animal Liberators: Research and Morality* (1988), Susan Sperling examines the animal rights movement as a social phenomenon, through the lens of a cultural anthropologist. She says animal rights proponents relate to animals as people, because they anthropomorphize their pets, the only animals they regularly see in their largely urban, industrialized existence. Sperling also maintains that the movement attracts people who are worried about technology run amuck in modern society, exploiting both animals and people. Her assessment is somewhat similar to that of the National Association for Biomedical Research, which says one "disturbing factor" that has helped animal rights philosophy gain adherents is "a basic anti-intellectual, anti-science sentiment that pervades today's society." The NABR concludes, "The public's scientific illiteracy and its perception of biomedical research as shrouded in mystery have helped ensure a favorable response to the anti-science cynicism underlying the animal rights position."

Regan, however, rejects the notion that his movement is anti-science. He calls this a "tired charge" that is a "moral smokescreen." The rights view is "far from being antiscientific," Regan says, because it "calls upon scientists to do science as they redirect the traditional practice of their several disciplines away from reliance on 'animal models' toward the development and use of nonanimal alternatives." Regan's comments reflect the views of a movement that wants to make modern society more compassionate. According to animal rights proponents, if humanity can overcome its speciesism, an effort that would entail a considerable restructuring of society, people would live in a more natural state, in harmony with the environment, its creatures and themselves.

Not all animal protection activists spell out their goals in such apocalyptic fashion. Many who express concern for animals do so in the framework of the more traditional humane societies dedicated to the immediate alleviation of unnecessary animal pain and suffering, not to the articulation of a completely new philosophy. These groups do not ground their activism in radical philosophy; they might argue that animal research to cure a deadly disease is acceptable, while it would be unjust to inflict pain or death on animals to invent a new brand of hair spray. Still, the all-encompassing philosophical arguments of rights proponents remain compelling to more and more people, and are making a mark on animal-oriented groups of every stripe.

Shareholder Resolutions from Animal Protection Activists

Despite the radical actions PETA supports, in some of its work on consumer product testing it has used tactics that are squarely within the tradition of much more moderate activists. In its use of shareholder resolutions, PETA has presented requests for the disclosure of animal use figures at consumer product companies. An independent shareholder, Linnea Pulis, has also raised the issue of animal testing at IBM, and the shareholder resolution process is now beginning to be used for other animal issues. PETA submitted its first resolution on selling fur coats in 1990, to American Express, and the ASPCA submitted proposals on factory farming to McDonald's in 1989 and Pepsico in 1990. Several groups proposed resolutions to U.S. Surgical in 1989 and 1990 on the company's use of dogs.

Organizing PETA's campaign: As noted above, in the mid-1980s, PETA began sending members to the annual meetings of consumer product manufacturers to demonstrate against use of animals in testing. The way was cleared for more substantive involvement in the meetings in 1985 when animal rights lawyer and activist Peter Lovenheim won an injunction that forced a company (Iroquois Brands) to include an animal rights resolution in its proxy statement, overruling the Securities and Exchange Commission staff view that the resolution did not pose a significant social issue. The issue that Lovenheim was raising had to do with the force feeding of geese, not product safety testing, but the decision appeared to set a precedent, and PETA's Ingrid Newkirk talked with Lovenheim about drawing up resolutions that would fit in with the product safety campaign.

Lovenheim drafted a resolution, and PETA bought between $1,000 and $2,000 worth of stock in Avon, Bristol-Myers, Gillette and Procter & Gamble so that it would have the standing to act as a shareholder proponent if need be (a proponent must hold at least $1,000 worth of stock in a company). PETA also inserted a notice about its planned campaign in its *PETA News* magazine that asked if members who held stock in any of a number of consumer products companies would be willing to submit the PETA resolution.

The first PETA member whose stock was used was Patricia Daley, a registered nurse and longtime stockholder in Armour-Dial, now part of Greyhound.

Daley, who is president of a child abuse prevention speakers bureau, had long been concerned about maltreatment of animals, which she sees as a precursor to child abuse. PETA has continued to look for members who will act as proponents. Subsequent PETA member proponents reflect the range of membership in the animal rights movement—a lawyer, a jewelry designer, a photographer, a former anesthetist, an educational consultant and a retiree. Most of them did not attend the annual meetings; Susan Rich of PETA appeared in their stead.

PETA's resolutions: The first resolution that Lovenheim drew up made it just under the filing deadline for Greyhound's May 1987 annual meeting. Like PETA's second resolution, which was filed for Procter & Gamble's annual meeting six months later, it was tougher than the proposal that PETA would submit for 1988 and 1989; instead of asking for a report on animal testing, it asked the company to halt testing not required by law and to begin to phase out products that are not marketable without animal tests. The original resolution did fairly well at Greyhound—6.7 percent support from the shares voted—but picked up only about 2 percent support at P&G.

After looking at the low P&G vote, PETA decided to tone down its proposal to a request for a report in hopes of appealing to a wider audience and improving its vote totals in the next proxy season. For spring 1988 annual meetings, it offered a new resolution to seven companies that asked for reports on progress in eliminating animal tests for consumer products and for a listing of products that are tested on animals in painful experiments. That resolution fared quite well at all seven companies: American Home Products, Avon, Bristol-Myers, Colgate-Palmolive, Gillette, Greyhound and Schering-Plough. But despite the change in strategy, the vote when the resolution was submitted to Procter & Gamble in the fall was still only about 2 percent, making it ineligible for reintroduction in 1990.

PETA returned to the same seven companies for the spring 1989 annual meetings and added a new one, Johnson & Johnson, after learning that the company was one of the largest users of dogs. Reflecting growing concern among companies about the tactics of animal rights activists, Johnson & Johnson hired a private investigator to check into the backgrounds of the two shareholder proponents, a move it later said it regretted.

PETA further softened the resolution in 1989 by eliminating the request for a listing of products that were tested in painful experiments. The 1989 votes were closely in line with the votes in 1988; the average was about the same, and the votes rose at four companies and dropped at three.

Attending annual meetings: In an interview with IRRC, Susan Rich of

PETA recounted her experiences with the 1989 shareholder campaign, when she attended six of the eight annual meetings where a PETA resolution was considered, missing only Colgate-Palmolive and Greyhound. That year, she said, the company officers treated the PETA representatives with respect, although the Avon meeting was "understandably very tense." (PETA was in the last stages of its boycott that Rich felt was causing a "public relations nightmare" at Avon.) Rich said that in earlier years she had been treated much more nervously at the various meetings; she cited being followed into the restroom at Gillette and stopped at a variety of checkpoints at Schering-Plough in 1988.

While most shareholder resolutions are disposed of quickly at annual meetings, Rich said that PETA carefully planned more extensive presentations for its resolutions by lining up a variety of speakers and that management made no effort to cut them off. She said that representatives of the Physicians Committee for Responsible Medicine spoke at Schering-Plough, Johnson & Johnson and Avon and that some discussions of the animal testing issue went on for more than half an hour. The best 1989 meeting from PETA's perspective, Rich said, was Bristol-Myers's relatively small meeting in Wilmington. She said management seemed particularly well-prepared to consider the issue, the discussion was thoughtful, and PETA got spontaneous support from unaffiliated shareholders.

While Rich was pleased with the attention given to the PETA resolutions at the meetings, she did not imply that they had been love feasts. She said it was PETA's intent to turn "the companies' show-and-tell into an unpleasant experience" for management. She also said that while the resolutions ask only for a report on progress toward eliminating tests, PETA's ultimate goal is to see that the companies stop using "even one animal."

Why no withdrawals? Both Rich and PETA lawyer Lovenheim expressed surprise that the companies that received product testing resolutions had not made overtures to meet with them to negotiate with an eye to providing the information that was requested in return for withdrawal of the resolutions. They indicated that they would certainly have entertained serious proposals for withdrawal, even though the requested reports were only a way-station toward PETA's ultimate goal of the elimination of all tests. Given the fact that most consumer products companies do appear to be making progress toward the elimination of animal tests, Rich said she found most of the companies' reasons for refusing to provide the reports (as enumerated in their proxy statement responses) to be distinctly "lame." Both Rich and Lovenheim contrasted the companies' attitudes toward the PETA resolutions with McDonald's willingness in 1989 to provide a report on factory farming in return for withdrawal of a shareholder resolution on that issue from the ASPCA.

The only withdrawal proposal PETA has gotten came from Procter & Gamble, and that was turned down, Lovenheim said, because "the company wasn't prepared to offer anything substantial." Rich speculated that perhaps word of that aborted negotiation had sent the other companies a message that "PETA was too tough or not serious about withdrawals." Lovenheim thinks the companies' attitudes on negotiations may change. "It's my sense that now that PETA has shown that it can get significant support and isn't going away, some of the companies will be more forthcoming," he told IRRC.

Looking ahead: PETA has continued its campaign into the 1990 proxy season. Rich points out that since 1990 will be the third year for the proposal at most companies, it will have to get 10 percent support in order to stay alive. Whether or not the resolutions get enough support for resubmission after 1990, Lovenheim thinks that dealing with the proposals has been "a valuable educational process" for management. He said he had found that corporate counsel and the corporate secretaries treated the proposals "cordially and seriously," by the 1989 proxy season, and he felt management had become "better educated about the subject as a result of the proposal process." Before PETA started proposing resolutions, Lovenheim pointed out, companies were exposed to the activists "largely through bumper stickers and placards. Now they're reading a carefully written 500 word statement, and it has led them to take the animal protection movement more seriously."

II. LAWS, REGULATORS AND LEGISLATORS

A. FEDERAL REGULATIONS

Federal agency policies on animal testing are unclear and contradictory where they concern consumer products that are not medical-related. This regulatory ambiguity means that both animal protection activists and industry officials can back up opposing arguments with substantiation from federal statutes, regulations and policy statements.

The safety testing of cosmetics and other nonmedical-related consumer products used in U.S. households mainly falls under the jurisdictions of the Food and Drug Administration and the Consumer Product Safety Commission, and to a certain extent the Environmental Protection Agency. Although there is no explicit animal testing requirement for these types of products, both agencies historically have used animal toxicity data as their standards. Both agencies also have elaborated on their policies on nonanimal alternatives in the last five years, opening the door to vastly differing interpretations of the actual requirements.

The Food and Drug Administration

The Food and Drug Administration has received more attention than any other federal regulatory agency on the subject of animal testing because of its authority over cosmetics. As animal protection activists are quick to point out, the FDA has no authority to require product testing for cosmetics. Companies are equally prompt to counter that if consumer safety is not substantiated before the product's distribution, the FDA has the authority to subject the company in question to regulatory action. In practice, this means that once a product is on the market and consumers complain to the FDA, the agency will conduct an investigation of the product. According to an FDA booklet that explains the agency's enforcement requirements, it can "inspect establishments in which products are manufacutured or held, and seize adulterated...or misbranded...cosmetics." If the company in question refuses to stop production of the offending product, the FDA may request a restraining order. Finally, the FDA may initiate criminal action against the offender, if all else fails.

Confusion arises because for substances other than cosmetics that are under FDA authority, regulations do rely on animal toxicity procedures to establish consumer safety standards. Further, when answering queries about its policy, the FDA is not always precise in defining which products it refers to, and those who ask questions of the agency are not always clear on which set of products they refer to. Unclear perceptions about the agency's regulations for different substances, combined with the FDA's "no, but..." answer to questions about testing requirements, have painted the murky picture of regulations now on display.

Each side in the polarized debate can produce a statement from FDA officials in recent years that supports a clear-cut position on testing, because at different times the FDA has said different things. Regulatory reality is much fuzzier than either camp would like to admit. The Federal Food, Drug and Cosmetic Act, administered by the FDA, prohibits the distribution of adulterated or misbranded cosmetic products. The booklet on the agency's requirements says, "A cosmetic is considered adulterated if it contains a substance which may make it harmful to consumers under customary conditions of use." The document continues:

> Although the FDC Act does not require that cosmetic manufacturers or marketers test their products for safety, the FDA strongly urges cosmetic manufacturers to conduct whatever toxicological or other tests are appropriate to substantiate the safety of the cosmetics. If the safety of a cosmetic is not adequately substantiated, the product may be considered misbranded and may be subject to regulatory action unless the label bears the following statement: Warning—the safety of this product has not been determined.

Although the statute does not give the FDA the authority to require any premarket testing of cosmetics, safety standards used by the agency when it reviews consumer complaints have relied on animal toxicity data in the past. So far, the agency has not accepted data from nonanimal alternatives when investigating product safety, according to a September 1988 letter to U.S. Rep.

Barbara Boxer (D-Calif.) from Hugh C. Cannon, the FDA associate commissioner for legislative affairs.

Since the FDA published its 1984 explanation, the agency has elaborated on its position, contradicting itself several times. In testimony to Congress on May 6, 1986, the acting director of the FDA's Center for Veterinary Medicine said the FDA "does not require the use of the LD50 to assess the safety of the products it regulates....Government agencies support the development and validation of alternative methods." He added, "the Draize test is still the most reliable method for determining the potential harmfulness, or safety, of a product instilled in the eye, such as ophthalmic drugs and devices and some cosmetic products." Alternative assays using cell cultures "cannot yet replace the Draize test," he concluded.

In a Feb. 29, 1988, letter to the Maryland legislature, FDA Commissioner Dr. Frank Young wrote: "The agency is aware that there are many potential nonanimal replacement tests which are in various stages of evolution but none have been accepted for such use by the scientific community at the present time. This response applies specifically to the Draize eye irritancy test and to all other acute toxicity tests of which the agency is aware." Referring specifically to shampoos and cosmetics, Young stated, "The responses and results of a tissue or cell culture test alone cannot, at the present time, be the basis for determining the safety of a substance." The commissioner concluded, "The FDA cannot permit the use of any potentially harmful substance in humans prior to preliminary testing in animals to provide reasonable assurance that it is not injurious to humans. Since certain tests should never be carried out in human beings and since at the present time there are no adequate alternatives, whole animal testing remains unavoidable."

Young's letter illustrates how confusion occurs about the FDA requirements. His statement about preliminary testing was in response to a question asking whether the Draize eye test is "necessary to establish the safety of cosmetic products under the regulatory control of the FDA." Young responded in the affirmative. Indeed, a Gillette official provided IRRC with a copy of Young's letter as proof that animal testing is required for cosmetics and household consumer goods produced by Gillette. The FDA has no authority to require any tests, however, and legally cannot "permit" anything until after consumer complaints about a specific product have been filed. Yet Young clearly is referring to premarket tests of cosmetics.

Precisely on this point, FDA associate commissioner Cannon wrote to Rep. Boxer in his September 1988 letter: "The Federal Food, Drug and Cosmetic Act does not give the FDA the authority to require cosmetic manufacturers or distributors to test their cosmetic products or ingredients for safety or make such data available to the FDA if tests have been conducted."

In comments to the Maryland Governor's Task Force on Animal Testing, the FDA seemed to encourage companies to use alternatives, while at the same time it did not say that it would necessarily accept data from these alternatives to substantiate product safety during an investigation. The April 17, 1989, statement reiterated that the FDA still considers the Draize test to be best and does not find that tissue or cell culture tests are sufficient to establish consumer safety. The agency did say, however, that "some in vitro studies are useful as screening tools to indicate relative toxicity."

The statement said the FDA "has provided industry with the following guidance relating to the use of in vitro methods to support the safety of products subject to eye irritancy testing:

1)   Because no one wishes to sacrifice animals unnecessarily and because the proper use of in vitro studies can reduce the total number of animals used in the development of a product, it is appropriate for industry to develop and use in vitro tests.

2)   Because of their inherent over-simplification of the physiology and response of the whole animal system, in vitro tests are not total replacements for the Draize eye irritancy test and probably can never be.

3)   A quick and inexpensive test, despite its inability to detect everything, can be used early on in the development phase of a product to eliminate chemical candidates that fail to pass, thus reducing the number of chemicals or formulations that would need to be tested eventually in animals.

4)   It is conceivable that in vitro tests, based in part on prior calibration with animal tests, could also be used as a final safety test in those situations where only a minor change in an inactive ingredient is made and where prior experience enables one to draw the scientific conclusion that the in vitro test is capable of detecting any likely changes due to reformulation.

5)   Since the FDA has no testing requirements for the premarketing of cosmetics, we have not developed testing protocols for that purpose. The four statements above represent our considered, informed opinion on the science. They should not be construed as regulatory [dicta].

Although there can be no official statements about what kind of tests the FDA requires to substantiate safety for cosmetics since there is no premarket clearance requirement for these products, companies naturally are interested in knowing what kind of data they may be asked to provide as safety substantiation if something does go wrong with one of their products after it is on the market. Even if this happens, though, the FDA cannot legally compel companies to show premarket testing data for the problem products. So far, animal toxicity data have been the standard for safety assessments, however. Periodic FDA comments on animal testing and alternatives, issued even though the agency has no authority to stipulate premarket tests, indicate what kind of safety standards the agency would use if it were to investigate a product after a complaint. Ultimately, an understandably conservative Catch-22 is perpetuated: The FDA will not say that it finds alternatives acceptable standards of proof until these tests are accepted by the scientific community and industry. Validation will come from further research of alternatives and eventually may be legally tested in a real complaint case that would probably stand trial in the courts. Yet corporate attorneys driven by the specter of consumer lawsuits are reluctant to approve nonanimal alternatives for premarket testing if there is no assurance from the FDA that it will accept these tests as proof of safety.

The Consumer Product Safety Commission

Confusion also surrounds the precise testing requirements of the Consumer Product Safety Commission for the substances under its jurisdiction: household products that are not medical-related. The Federal Hazardous Substances Act, which covers these products, has clearer language in its requirements than the Federal Food, Drug and Cosmetic Act's broad charge on cosmetics, however. The CPSC also has issued an official animal testing policy, in contrast to the FDA's many obfuscating statements. Still, it is not clear if this policy, published in the *Federal Register* in May 1984, supplants the CPSC's earlier regulations or simply modifies them to some unspecified extent.

Regulations under the Federal Hazardous Substances Act provide for the proper labeling of hazardous products. They define a material as hazardous "if such substance or mixture of substances may cause substantial personal injury or substantial illness during or as a proximate result of any customary or reasonably foreseeable handling or use, including reasonably foreseeable ingestion by children." The only hazard category to use animal toxicity data in the initial section is the "highly toxic" ranking. Although human data, if available, are to take precedence over animal data, the statute uses the following definition: A substance is highly toxic if it a) kills half the test population of at least 10 rats within two weeks after they have been fed the material; b) kills half the test population of at least 10 rats within two weeks after they have inhaled the material for one hour or less; or c) kills half the test population of at least 10 rabbits within two weeks after the material has been in contact with their shaved skin for 24 hours or less. Animal toxicity standards do not appear in the other hazard categories of this initial section. The regulations do refer to another part of the Federal Hazardous Substances Act that sets forth in detail the proper method to test toxic substances, though. This section describes rabbit tests for determining acute dermal toxicity, primary skin irritancy and eye irritancy.

An alternative definition for "highly toxic" is given later in the regulations, again stating that, if available, human experience outweighs animal data. The alternative definition describes two more categories using animal data: It says animal tests similar to those described under the "highly toxic" classification above can define "toxic" substances, while a Draize skin irritation test on rabbits can define "corrosive" materials; still, human data should be used if they exist. The main difference in the alternative definitions comes where they specify only that "The number of animals tested shall be sufficient to give a statistically significant result and shall be in conformity with good pharmacological practices," instead of saying how many animals must be used in the tests.

Despite the clear language in the regulations, companies and activists do not concur on the CPSC's stance. They have reason to disagree. In its May 1984 animal testing policy statement, the CPSC stressed,

> [I]t is important to keep in mind that neither the [Federal Hazardous Substances Act] nor the commission's regulations require any firm to perform animal tests. This statute and its implementing regulations only require that a product be labeled to reflect the hazards associated with that product. While animal testing may be necessary in some cases, commission policy supports limiting such tests to the lowest feasible number and taking every feasible step to eliminate or reduce the pain or discomfort that can be associated with such tests.

The commission then urged the use of alternatives, but did not include in vitro tests in its list of options:

> [T]he commission and manufacturers of products subject to the FHSA should whenever possible utilize existing alternatives to conducting animal testing. These include prior human experience, literature sources which record prior animal testing or limited human tests, and expert opinion. The commission resorts to animal testing only when the other information sources have been exhausted.

Specifically, where tests are still necessary, the CPSC recommended replacing the LD50 with a limit test that uses 10 to 20 animals instead of 80 to 100; not testing for eye irritancy if the substance is a known primary skin irritant; using an ophthalmic anesthetic before placing the substance in rabbits' eyes; and using a tiered testing approach to reduce the number of animals, by gradually increasing the dose for animals until irritation is determined instead of using a larger number of animals for initial tests. Also, rabbits in skin irritancy tests should not be placed in stocks and should have access to food and water while the substance is on their skin, the policy says.

The 1984 document concluded by referring to nonanimal alternatives:

> When such techniques are accepted by the scientific community as adjuncts or alternatives to whole-animal testing, [the CPSC's] Health Sciences [Laboratory] will incorporate the techniques into the commission's compliance program to the extent feasible and will recommend any changes to the commission's statutes or regulations that may become appropriate as the result of advances in testing methods that are developed.

Both sides in the debate dispute whether the 1984 policy replaces the regulations and disagree about what exactly is implied by the commission's stress on not requiring animal tests. This uncertainty is furthered by statements the CPSC has made. In a March 1988 letter to a Maryland legislator, the CPSC's Associate Executive Director for Health Sciences, Andrew Ulsamer, wrote that although the statute does not "expressly require product testing on animals," in some instances the alternative methods listed by the commission's policy are not adequate and animals must be used. "At the present time we do not believe that an adequate nonanimal replacement exists for either the Draize eye irritation test or other acute toxicity tests. The nonanimal tests presently under development are not yet at a stage where they can be validated prior to their incorporation into regulatory testing protocols," Ulsamer said. Consequently, "when existing animal test data, human experience or expert knowledge is inadequate to determine toxicity, acute toxicity tests such as [the] Draize eye irritation tests are necessary to facilitate the appropriate labeling of hazardous consumer products." These comments seem to reveal that the CPSC's public emphasis on not requiring animal tests has little meaning in practice, since the commission acknowledges that the only standard it really will accept comes from the animal data it says are not required—despite its listing of some alternatives in the 1984 policy statement.

A statement made to the Maryland legislature in March 1989 by Dr. Edward Jackson, director of toxicology and research services at Noxell Corp., illustrates how companies interpret the commission's semantic gymnastics. While the Maryland legislature was considering a bill to ban the use of acute toxicity tests on live animals within the state, Jackson stated, "for household products, passage of [the bill] would be in direct opposition to the Federal Hazardous Substances Act regulations." A January 1989 memo on federal policy to the Maryland Governor's Task Force on Animal Testing concludes, "It is clear from [Jackson's comment] and other statements made by industry that the animal testing policy advanced by the Consumer Product Safety Commission, which embraces the use of alternatives to animal testing, has not been translated into a regulatory framework which supports this policy determination."

Compared with the FDA, the CPSC has more authority to require specific tests to determine product safety. Until 1984, regulations using animal data to define "highly toxic" substances implicitly meant that companies had to test on animals to label their products properly. After the CPSC issued its policy statement in the *Federal Register* in 1984, however, requirements became murkier. Currently the CPSC seems to be using the same approach as the FDA, to an extent, by emphasizing that it does not require animal tests but at the same time saying that it does not find nonanimal alternatives an acceptable measure of consumer safety. The difference between the two regulators is that Federal Hazardous Substance Act regulations still on the books continue to define some toxicity levels in terms of animals. Until the CPSC makes an explicit regulatory announcement modifying its current rules and approving at least some nonanimal alternatives, many companies certainly are not likely to take a leap of faith and adopt only nonanimal methods for household product testing. The same cycle in place at the FDA that keeps much of the cosmetics industry testing on animals seems to maintain the status quo for many consumer products companies.

Other Federal Agencies

The FDA and other federal agencies also use data from animal tests to determine the hazards posed by various substances not so directly related to consumer products:

- In addition to its regulation of cosmetics, the FDA also has jurisdiction over food, drugs, medical devices, color additives and radiological products.
- The *Environmental Protection Agency* uses animal data to determine the toxicity of pesticides, industrial chemicals, air and water pollutants, hazardous wastes and some radiation hazards. (In fall 1988 EPA modified its position on animal testing and said it would accept some test modifications to the LD50 that use fewer animals than the classic method.)
- *Department of Labor* regulations say that dangerous workplace substances must be identified, and most tests that are currently used to iden-

tify these substances use animals.

- Animal tests also are used to categorize the different toxicity levels of hazardous materials shipped in interstate commerce under *Department of Transportation* rules.
- The *Department of Agriculture* conducts some animal tests to determine food safety.
- The *Centers for Disease Control* use animal data to keep track of agents that could spread communicable diseases.
- The *Federal Trade Commission* reviews animal testing data to substantiate the packing and advertising claims of companies.

In addition, the Animal and Plant Health Inspection Service (APHIS) of the Department of Agriculture administers the annual reporting on animal use required under the Animal Welfare Act by research facilities that conduct animal tests.

The Animal Welfare Act regulates housing, feeding and personnel training for laboratory care in facilities that conduct experiments on animals (excluding mice, rats, birds and farm animals). The act became law in 1966 and was amended in 1970, 1976 and 1985. Each facility is required to have a committee to oversee animal care. The FDA, CPSC and EPA also set Good Laboratory Practices standards for toxicity data; the congressional Office of Technology Assessment reported in a 1986 study of animal use that "because proper animal care is essential to good animal tests, these rules indirectly benefit animals."

The main criticisms of federal oversight of animal welfare at laboratories, according to the OTA study, are that Good Laboratory Practices guidelines do not ensure use of pain relieving drugs; inspection of facilities is insufficient; the most common laboratory animals—mice and rats—are not protected; and the law does not regulate the design or performance of actual tests. The 1985 amendments to the Animal Welfare Act, which require exercise for dogs, larger cages for animals, and provisions to care for the psychological well being of primates, have been praised by animal advocates, but some researchers contend that the cost of the new requirements will be prohibitive.

B. FEDERAL AND STATE LEGISLATION

Animal protection activists have instigated legislation at both the federal and state levels. The proposed federal measure would require regulatory agencies not to accept LD50 data and to review their acceptance of other data based on animal tests, while proposed state legislation has focused on amending existing laws that forbid cruelty to animals, by amending the definition of "cruel" to include toxicity tests. Strong opposition from the biomedical research community and industry groups has blocked the passage of all proposed restrictions on animal testing, but both activists and their opponents have been surprised at the strength of support for these restrictions. Because the debate over legislation sets forth the arguments in the controversy over animal use for product testing and has involved nearly all the players concerned with the issue, it offers a good summary of the positions of the warring factions.

The Proposed Federal Bill

The proposed law: The Consumer Products Safe Testing Act, now pending for a fourth year on Capitol Hill, is the only piece of proposed federal legislation to address directly the issue of animal toxicity tests for consumer products.

In its introduction, the bill says that the LD50 test is "inaccurate, misleading and unnecessary in product testing"; that other tests are "less costly, more humane and more accurate, [and] nonanimal alternatives have been developed for other acute toxicity tests using animals"; that the federal government encourages the use of the LD50 and other animal tests; that private industry "is reluctant to use other tests without encouragement" from the federal government; and that industry and consumers "will benefit from the promotion of alternative methods of testing when these alternatives are more accurate and humane than animal tests."

There are two main provisions to the bill. First, it specifically prohibits federal agencies from using the LD50 "when determining product safety, labeling or transportation requirements for the purpose of federal regulation."

Second, the proposed measure would require that nonanimal tests be used instead of animal tests, except where justification for animal use is provided. The bill says the heads of federal agencies must "review and evaluate" their existing positions on animal testing and "promulgate regulations which specify that nonanimal tests be used instead of animal toxicity tests, unless that federal department or agency head determines that in certain limited cases the nonanimal toxicity test has less validity than the animal toxicity test." After this review, if the head of an agency finds a nonanimal test "less valid than an animal toxicity test," a justification for the continued use of animal tests must appear in the *Federal Register*, open to public comment. Further, the continued use of animal tests is to be reviewed and justified every two years, with a report in the *Federal Register* subject to public comment.

Holly Hazard, an attorney who helped to write the legislation, told IRRC that the bill's definitional section is important to understanding its intent. Hazard, who is also executive director of the Doris Day Animal League, pointed out

that the animal tests that would require review and justification are listed in
the bill's definitions as the Draize eye and skin irritancy tests, two commonly
used variations of the LD50--the approximate lethal dose and the limit tests,
and other unspecified acute toxicity tests that use animals. This list covers
most of the range of tests performed for consumer products, but some of these
tests are also used for biomedical research.

Venturing into the realm of medical research would stretch the debate beyond
the issues that apply only to testing of consumer products. Hazard said,
however, that the bill's definitions make clear that the bill does not cover
medical research. The definition of federal agencies to be affected by the bill
is important to understanding its meaning, she maintained. That definition
applies the bill to agencies that have the "authority to promulgate regulations,
guidelines and recommendations with respect to procedures to be used in the
testing of products, including consumer products, veterinary products and
products containing hazardous or toxic substances." Agencies that license or
approve "products, labeling requirements for products, or the transportation
of products based on the results of these tests" are also covered. This means
that medical research is exempt from the bill, Hazard said, although the
products covered go beyond cosmetics and household cleaners to include
substances such as pesticides. Hazard's assertion may be open to challenges;
the phrase "products containing hazardous or toxic substances" is broad
enough to encompass pharmaceutical products, and could be interpreted to
mean other sorts of medical devices as well.

Rep. Barbara Boxer (D-Calif.) introduced bill HR 1676 in the House, where it
picked up more than 80 cosponsors in 1989; the identical Senate version (S 891)
was introduced by Sen. Harry Reid (D-Nev.). In 1988, the House version
ended up in the Energy and Commerce Subcommittee on Health and the
Environment, whose chairman is Rep. Henry Waxman (D-Calif.); on May 16,
1988, Waxman's subcommittee heard testimony, but it took no further action.
By January 1990 the House version still had not progressed, but a hearing on
the Senate side was held on Nov. 8, 1989.

Debate on the Bill

Representatives from both sides of the animal testing debate crossed verbal
swords during the May 1988 hearings on the bill, expressing widely divergent
opinions on what role the federal government should play in regulating animal testing. Their testimony clearly illustrates the lack of common ground.

Discussion focused on the LD50—which the bill would ban specifically, and
on the controversial Draize irritancy tests—which would be covered by the
review and justification provision; animal tests in general also were discussed.

Although supporters said the bill would not affect medical research, its language is ambiguous, as seen above. Opponents seemed to feel the entire biomedical research establishment was under attack and usually did not differentiate between consumer product tests and medical research—perhaps because they view limitations on use of animals in product safety testing as a foot in the door for limitations on medical research. They construed the proposed LD50 restriction and toxicity test review process as but the first in a possible string of anti-science measures, and predicted dire consequences for the country if the bill were to become law.

Although industry groups strongly object to proposed restrictions on any kind of animal testing, the federal hearings did not include statements from two groups that have figured prominently in state hearing rooms: the Cosmetic, Toiletry and Fragrance Association and the Soap and Detergent Association. In the mainstream corporate establishment touched by the debate, these two associations have supported the most research on possible alternative tests that do not use animals. Their absence from the federal hearings is somewhat puzzling, given their prominence in the debate overall. In a conversation with IRRC, Holly Hazard suggested that these two industry groups might have difficulty reconciling their state testimony—that federal regulations constrained companies' adoption of alternatives—with federal testimony against the same "restricting regulations." Representatives of the two groups would not elaborate to IRRC on their reasons for choosing not to testify at either the House or the later Senate hearing.

Supporters of the bill berated what they saw as federal inaction and said that alternatives are available now to replace animal tests. The bill's backers described toxicity tests now in use, said they were cruel, and testified that the pain and suffering experienced by animals in these methods is not justified. Focusing on consumer products and touching on medical-related testing, they argued that the bill would influence government regulators and companies toward positive change and promote the constructive use of alternatives, reducing animal use considerably.

Position of the proponents: A common theme voiced by the bill's proponents was that, on their own, federal agencies would not make a decisive move to encourage alternatives. "It seems the regulatory environment is not receptive to alternatives. I think the Congress can break this inertia," Rep. Boxer said at the outset of the hearing. "It is not enough to just get rid of the classical LD50—we must ensure that we have in place a regulatory mechanism which clearly demonstrates to industry that the U.S. government will consider nonanimal alternatives for testing products, and will ensure that data from such tests will be accepted if the tests prove the equal of animal methods," Boxer said. "The sluggish nature of federal reform has led to a policy with respect to testing procedures which is completely out of touch with modern scientific advances," Hazard testified.

Hazard felt that companies would be prodded into using more alternatives if regulatory ambiguities were cleared up, and she said the bill would resolve the confusion about federal requirements that has stymied progress. However, she warned that, "If industry and scientists continue to believe that innovative testing techniques will be rejected by the agencies from which they must gain approval for marketing their products, research and testing in this area will stop." The bill would put nonanimal tests on an "equal footing" with animal tests, Hazard said, maintaining that because such tests do not have this status in the regulations now, it is "understandable" that companies see animal tests as necessary.

Two toxicologists who spoke in favor of the bill outlined what they described as viable alternatives that do not use animals. Although neither said that all animal tests could be replaced completely by currently available methods, neither did they give the government many plaudits for its role in encouraging alternatives. Dr. Christopher Kelly, head of the National Testing Corp., which developed the Eytex method to replace the Draize test for a range of products, described his test as one of the available alternatives. Eytex has been validated by 16 private and federal labs for up to 60 substances at each lab, and high correlations were shown with Draize test values, Kelly testified. He cautioned, however, that "A small number of substances are not qualified in the [Eytex] method at any meaningful concentration, and substances which irritate the eye by physical mechanisms must be tested in vivo [on animals]."

Kelly said his test had broad potential. He placed blame for the lack of its wide acceptance squarely on regulators' shoulders: "The market growth of the [Eytex] system has been hampered by inadequate and inconsistent standards set by government regulators to determine product safety. Current federal regulations limit the advancement of product testing techniques." The bill would make federal agencies review their regulations, statements and guidelines to "create clear and equitable performance standards for innovative and improved methods," Kelly said. His final point was that industry use of alternatives such as Eytex "can help protect industry from unfair regulations in reaction to the growing opposition to animal suffering."

Another toxicologist in favor of the bill was Dr. Gerhard Zbinden, formerly head of toxicology at Hoffmann-La Roche, now director of the Institut fur Toxikologie at the Universitat Zurich. Zbinden said that he is not unalterably opposed to animal tests, although he feels that the lowest possible number of animals should be used. He estimated that about 80 percent of all biomedical research could be done with in vitro methods. The LD50 is a "classical example of an originally good idea gone astray," he said, and listed its major shortcomings: Different laboratories produce different LD50 values, and the test provides information only on the lethal effects of a substance, with nothing on its non-lethal effects. Also, many animals "suffer great pain and anxiety," even though the information they provide "is of minor practical and clinical significance." Zbinden concluded, "Inertia and excessive reliance on a test procedure [the LD50] that has enjoyed almost unquestionable prominence over many years has made it difficult to place the assessment of acute toxicity on a radically new basis."

A number of physicians who treat accidental poisoning victims also submitted statements in support of the bill. These doctors said they had never used data from animal tests such as the LD50 or Draize to treat their patients because they prefer clinical data from humans. Dr. Murray Cohen, chairman of the Medical Research Modernization Committee, an animal protection group, said in a personal statement that drug dosages recommended by drug manufacturers based on LD50 values were inaccurate—not enough to be effective in some cases, or too much in other cases. He concluded:

> This trend for pharmaceutical houses to consistently recommend subtherapeutic doses of useful drugs reflects their conservatism and their understandable desire to avoid litigation. Unfortunately, their use of the LD50 has been woven into this need for legal protection, and, whereas it may irrationally provide drug houses with protection, it results in inadequate care for sick patients.

Other animal protection activists have brought up the liability protection issue in relation to consumer products as well, expressing similar sentiments. Cohen said the LD50 is

> a hopelessly outdated, notoriously cruel, medically insufficient, and totally unnecessary test, and should be prohibited....The reasons for this conclusion include non-extrapolatability of animal data to humans, inconsistent results, the ignoring of sublethal effects, the ignoring of chronic toxicity, the ignoring of all biological systems save the most sensitive one, valuelessness in toxicological emergencies, and worthlessness in drug interactions.

Boxer concluded, "Claims that this legislation will force all animal testing to a halt are patently misleading and irresponsible. The bill is directed at the classical LD50 in the first part, and in the second part we are saying, if you must use animal tests, please justify it." She said that "a test of society is the way it treats all living things. I urge the committee members to allow all of us to measure up to that test."

Position of the opponents: Most of the bill's opponents concentrated on describing the effects they felt the bill would have on biomedical research, and they generally did not directly address the issue of nonanimal tests for consumer products—the current target of activists at major U.S. companies and the ostensible focus of the bill. The American Physiological Society was the only witness in opposition to submit testimony precisely on the consumer product issue. The Society said the proposed bill does not address just consumer products, but "also would affect some of the testing methodologies used in biomedical research." The Society wrote that "it is difficult to separate the testing of cosmetics and household products from medical and pharmaceutical testing." This is because "ingredients used in the formulations of cosmetics are similar to those used for the application of drugs to treat skin or other diseases....The general cosmetic base is used [because] the base has low-irritancy, non-toxic and emollient properties that support the pharmaceutical agents."

The Society said the bill's requirement for public justification of all animal tests would create "bureaucratic delays" that "would halt research efforts" in search of cures for ailments ranging from "cancer and AIDS to acute and chronic pain." Finally, it said, supporters of the bill "do not appreciate the complexities of toxicity testing nor understand the relationships between consumer product testing and biomedical research in assuring the public's continued state of well-being."

Dr. David Rall, director of the National Institute of Environmental Health Sciences, was one of the bill's most fervent detractors. He predicted "grave consequences for the nation's health" if the measure became law. The "scientifically unfounded prohibition and requirements [proposed] set a very disturbing precedent for all biomedical research," Rall found. "The implication that there are nonanimal models for toxicity tests currently available to provide adequate information for protection of the public health and safety is not true," he added. Alternatives are being developed, but they are not ready for implementation, he said, and although some modifications to the LD50 are acceptable, sometimes information from the classical LD50 is needed. Passage of the measure "would have devastating repercussions for the nation's health," because "even with the promise that alternative methods hold, I do not foresee a day when they will, in any comprehensive fashion, replace whole animal models." Rall presented a list of 60 "national and voluntary health organizations" that also opposed the bill, and a resolution from the National Cancer Advisory Board supporting continued medical research using animals.

In a response to a question from Waxman, Rall acknowledged that the National Institutes of Health have not used the classical LD50 since the early

1980s. He said a modified LD50 might be necessary in determining a combined drug therapy for treatment of certain cancers, but he did not mention any need for the LD50 in respect to non-medical consumer products. His statement that information from a precise LD50 might sometimes be needed was backed up by a letter from the American Medical Association's executive vice president, Dr. James H. Sammons, who said in part:

> The use of the LD50 remains crucial to the medical profession. The LD50 is essential for the characterization of the toxic effect of drugs and thus assists in determining harmful effects on patients....The determination of a statistically significant LD50 may require the use of up to 50 rats or mice. Other statistical techniques are available which require fewer animals. However, these methods do not provide confidence limits or complete dose response information. This is not an unimportant matter since these values are used to develop the margin of safety and therapeutic index of a drug for human use.

David Blake, associate dean for research at Johns Hopkins University, testified that the bill makes no distinction between the classical LD50 and the validated modifications of the test and does not provide for exemptions if the LD50 were needed in the cases that Rall mentioned. Blake was representing the National Association for Biomedical Research, a group funded by institutions and companies connected to biomedical research. He also said that "a few foreign countries still require the precise LD50 test for marketing products, and the European Economic Council requires exact LD50 values for transporting chemicals in some situations."

All of the bill's opponents echoed Rall's sentiment that although research continues on alternatives to the use of animals, no nonanimal tests have been validated yet. Opponents were quick to point out that although they did not believe animal tests could be completely replaced, they did accept as valid some modifications to the tests, which use fewer animals and administer some painkillers to the animals used. John Moore, assistant administrator for the office of pesticides and toxic substances at the Environmental Protection Agency, pointed out that EPA had modified its requirements for the classical LD50 in 1984 by saying that it would accept data from the limit test, for example, which uses six to 10 rats or mice, instead of the classical version's 60 to 200 animals.

Legislation cannot force science, Blake said in his testimony: "Only science drives the selection of the best means of safeguarding public health and safety and preventing exposure to unknown risks. The passage of laws or regulations determining the exact safety tests to be employed may only frustrate development of innovative methods." He concluded, "Science is responding as fast as it can. We trust the public will continue to closely monitor our progress."

Analysis: Congress is left to make sense of the issue, using conflicting testimony from a complex debate. One of the main points at issue concerns what kind of animal testing will be affected by the bill. Only one of the many opponents that testified, the American Physiological Society, mentioned a connection between consumer product testing and medical testing—a connection that animal advocates say need not be made. As discussed above, proponent Hazard explained to IRRC that the bill was intended to cover all product testing, not just consumer product testing, but that medical research was exempted by the bill's definitions. The mention of "consumer products" in its title was included to highlight to the public that tests of products used in everyday living involve animal cruelty, she said. Opponents of the proposed measure do not agree with this assessment and insist that biomedical research would be hurt badly if the bill became law.

Also under consideration is whether alternatives that now exist are acceptable replacements for animal use. Animal activists say yes; their opponents hotly deny it. Semantics confuse this part of the debate: Traditionally, "alternative" has meant one of the "3 Rs"—refinements of tests (reducing pain), reduction in the number of animals used and replacement of animals. But now many animal rights activists are insisting that replacement is the only moral course. Animal welfare proponents are willing to settle for refinement and reduction in the short term. Companies and government officials say they are using alternatives when they mean refinement and reduction, but not necessarily replacement. Progress on replacement alternatives—involving in vitro tests, computer modeling and other nonanimal methods—has bloomed in the last decade, and it seems to be accelerating. Finally, however, although company and government toxicologists have implemented a variety of refinements and reductions, they are not yet prepared to adopt a battery of alternatives to replace all the old tests.

The regulatory inertia attacked by activists is quite real, but the activists' characterization of their opponents as sadistic animal murderers is political theater, effective though it may be in obtaining media coverage. Weight must be given to the physicians who spoke in favor of the bill who say that they do not use animal data, but also at issue are the reasons why all the main medical professional associations spoke against it.

The biomedical research community is extremely wary of any government restrictions on how it conducts its business, while the animal protection movement would like to see animal tests eventually forced out of existence, by legislation or other public pressures. Consumer products are the first on animal advocates' agendas, but they are not the only thing on the ultimate wish list of many activists. Congress has not come to any conclusions, and it is not clear if the bill will move to a floor vote, or remain stalled in committee. Further progress on federal legislation could come from redrafting the measure, or could be instigated by scientific breakthroughs. Barbara Pequet, director of governmental affairs for the ASPCA, told IRRC that although no reactions to the mid-1989 decisions by Avon and Revlon to stop animal tests were apparent yet in Congress, those companies' actions "could be very pivotal" to the future of the federal bill.

Proposed State Legislation

State action has focused on amending existing laws that forbid cruelty to animals to define toxicity tests as unnecessarily cruel so that the tests will be illegal under current law. Bills on animal toxicity tests were proposed in seven states during 1989. (See chart for specifics in each state.) Although each of the bills differed somewhat, in general they proposed banning tests that use animals in eye and skin irritancy tests and acute toxicity tests for cosmetics and household products. Some of the bills affected only skin and eye irritancy tests, in which rabbits are commonly used, while others focused on acute toxicity tests as well. Bills in Maryland and Massachusetts would have limited the ban to cosmetics, while the other states would have included household products. Four of the pieces of proposed state legislation specifically exempted medical research from the potential test ban, and the Maryland version would have allowed research on the effects of cosmetics poisoning. The most stringent proposed penalty was in a California bill,which would have made some violations of the proposed law a felony (violations in other states would be only misdemeanors). In a twist, a New Jersey measure would have banned all eye and skin irritancy tests unless in vitro methods were used as well.

Proposed 1989 state legislation on laboratory animals was not confined to testing restrictions. Several measures would have restricted the use of animals from pounds for research, and a proposed California law would have required labels identifying products that were tested on live animals. Further, in Louisiana and Minnesota, legislatures approved bills designed to discourage laboratory break-ins, while in California and Wisconsin similar measures were received favorably in committee.

The Maryland experience: A closer look at one of the state debates illustrates how grassroots animal protection activism has percolated into legislatures around the country. Although a 1986 New Jersey bill proposed an animal testing ban, in 1987 Maryland became the first state where a bill specifically targeting product testing made any headway. The group that has

Proposed State Legislation on Animal Testing, 1989

State	Bill's Provisions	Status
California A.B.2461	Make use of Draize or LD50 illegal	killed after hearing in June 1989
Connecticut S.B.394	ban Draize test	rejected in committee
Hawaii S.B.332	ban LD50 and Draize for cosmetics and household products	rejected in committee
Illinois H.B.1042 S.B.1061	ban skin and eye irritancy tests for cosmetics, household products, washing, cleaning and laundry products	House: interim study by committee; Senate: killed in committee
Maryland S.B.654	ban eye irritancy and acute toxicity tests for cosmetics; allow tests for medical and cosmetics poisoning research; require report on animals tested in State	died in committee after hearing in March 1989
Massachusetts S.B.52	ban eye irritancy and lethal dose test for cosmetics products	stalled after hearing in March 1989
New Jersey S.B.109	ban Draize eye and skin tests for all products and medical research unless in vitro testing is also used	passed Senate, did not pass Assembly
Pennsylvania H.B.873	outlaw eye irritancy and acute toxicity tests for cosmetics, household products; various other animal-related provisions	hearing in May 1989

pushed the Maryland bill, Maryland Legislation for Animal Welfare, was formed in 1986 by Diane Nixon and Frank Branchini, former volunteers with People for the Ethical Treatment of Animals who had attended meetings PETA held to spur interest in work on animal-related legislation. Branchini told IRRC that he and Nixon initially tried to work through the Maryland Federation of Humane Organizations, a coalition of about eight statewide groups. Cooperation did not work out, Branchini said, so he and Nixon helped to create their own group, "with the idea that all humane organizations in the state would be interested and supportive of our efforts." This assumption "played out positively," Branchini told IRRC, and now most animal protection groups in the state are involved "in some way or another."

Maryland State Sen. IdaMae Garrott, who had worked on animal-related legislation in the past, met with interested activists while Maryland Legislation for Animal Welfare was forming and asked for more support for her efforts at the Capitol in Annapolis. Until this time, support for humane bills had been limited mostly to testimony at hearings. As a first step, the humane groups set up phone trees to jam key legislators' phone lines when animal issues came up. Cooperation on this effort began to build grassroots support for the testing bill, but help really started to pour in when PETA asked all its Maryland members to work on the issue, Branchini said. Initially 500 activists pledged their support to Maryland Legislation for Animal Welfare; this number grew to the current 3,000 members. Branchini stressed to IRRC that these 3,000 are not just occasional volunteers, but committed activists who regularly devote time to the issue.

Sen. Garrott first introduced the legislation in February 1987, proposing a simple ban on eye irritancy tests (including the Draize) and acute toxicity tests for cosmetics and household products. Somewhat different versions of the 1987 bill passed both the House and Senate by wide margins, then went to a conference committee for resolution and deadlocked in a tie. Despite revisions, the bill died in committee the following two years, as well. The 1989 version was twice as long as its 1987 predecessor. In addition to the initial proposed ban on certain tests, this most recent version would have animal testing facilities establish written animal care standards and guidelines and would require testing facilities to file detailed yearly reports on their animal use. A final new provision would explicitly exempt medical research from the ban and also allow for "the use of live animals to develop antidotes for ingestion of or exposure to cosmetic products."

Progress for the testing ban bill during the 1990 legislative session in Maryland is uncertain. Branchini said his group is in a "strange position...because committee changes were not good for us." Nevertheless, he said, the vote in 1989 was very close and there is hope for the bill's passage.

Despite the support Branchini described, strong opposition from both the biomedical research community and various industry groups continues to block the bill's passage. The University of Maryland, area poison centers and major Maryland newspapers also voiced opposition to the measure. As in other states, the Cosmetic, Toiletry and Fragrance Association testified that its member companies have a moral and legal responsibility to test their products on animals because no alternatives are acceptable as yet, and companies need to ensure that their products are safe for people. Further, animal testing data are needed if consumers are accidentally poisoned, CTFA said. The industry group also described its commitment to funding research into alternatives and to minimizing the use and discomfort of animals until alternatives are found.

A common argument mixed into testimony against the current bills that target only consumer products is that any test prohibitions will affect medical research. Sentiment seems to be that if even one bill passes to constrain animal use in product testing, dozens of tougher measures will follow. Troy Soos, president of the Alliance for Animals in Massachusetts, told IRRC that this is a "Saturday Night Special argument," and that once legislators have worked to produce a bill on any issue, he does not think they are inclined to revisit the subject again soon. Branchini, as well, noted that it is harder to attack testing for products that are not seen to be as frivolous as cosmetics, and that any testing bans that "smack of medical research won't pass." Still, rights groups do not deny that they would like to stop all animal use.

Task force established—In action related to the proposed Maryland bill, Gov. William Donald Schaefer (D) established the Maryland Governor's Task Force on Animal Testing in January 1989, to "study and make recommendations for the humane treatment of animals used in the testing of cosmetic and household products." After six months of meetings with the public, industry groups and animal advocates, the task force concluded on July 1, 1989, that animal testing for these products is still necessary and that a bill banning product tests that use animals should not be supported. The task force noted that "substantial progress" has been made on developing alternatives that do not use whole animals. Its recommendations: 1) Maryland should support businesses that use alternatives and attract companies that are developing alternatives; 2) an Animal Testing Council from the public and private sectors should be created to monitor the issue; 3) federal regulatory agencies "should be encouraged to clarify their position on the use of alternatives."

The 13 members of the group included representatives from three departments of the Maryland state government, three citizen representatives, two state legislators, an attorney, and representatives from Noxell Corp., Baltimore Animal Control, the University of Maryland and the Johns Hopkins Center for Alternatives to Animal Testing. Animal protection groups in Maryland were not happy with the task force, since they felt it was stacked against them. Still, they did not necessarily feel the report would have completely negative effects, since it did examine both sides of the debate. One issue the report looked at was whether a test ban would discourage companies from coming to Maryland. Although it concluded that the business climate might be adversely affected if the proposed bill were to pass, Branchini did not find this of concern. The idea that the bill would "send the wrong message to business is perhaps a valid concern, but none of the legislators will tell constituents that is why they voted against the bill—such arguments won't convince the public," Branchini told IRRC.

Flood of grassroots support, spread to other state bills—Despite the standstill in Maryland, animal protection activists have been sanguine about their progress, as in other states. Hazard of the Doris Day Animal League told IRRC that the 1987 Maryland bill "went so far so fast that it stunned the research community and animal rights activists." Intensive lobbying took place on both sides of the issue: Maryland legislators on the committee considering the measure had to contend with two lobbyists each, Branchini said, adding that, "although this may be common in Washington, it's really unheard of in Annapolis." There has been "unprecedented support" for the bill, he said. Activists have identified their legislators by district and committee, and are ready to activate phone trees at short notice. When the bill was considered in the conference committee, for example, the state assembly's switchboard reportedly had to be disconnected because it was besieged by so many callers.

The Maryland experience has been repeated in some variation in the other states considering bills on animal testing. Common to all efforts is a strong local component, although animal groups have shared information on their experiences. For instance, Branchini told IRRC that the Maryland Legislation for Animal Welfare office was flooded with inquiries once word of its efforts spread through the United States to other activists. The Maryland group eventually put together a legislative "how to" packet and shipped out several hundred copies around the country. Some national coordination also has come from conferences in Washington, D.C., sponsored by the National Alliance for Animal Legislation. The trend, though, has been for strong local organizations already working on animal issues simply to include lobbying on the testing bills in their current agendas, or to spin off new groups with a legislative focus.

Assessment of state efforts: Despite having posted no clear victories in their efforts to pass legislation banning toxicity tests for cosmetics and household products, most animal protection activists do not seem daunted. One exception may be in Connecticut; Julie Lewin, a lobbyist for the Fund for Animals, explained to IRRC that a number of large companies that test on animals have headquarters in the state and have influence in the legislature that will probably block any progress. Others are more optimistic that their efforts are having some effect. Sylka Morrison of Illinois Citizens for Humane Legislation noted that her group had been "dismayed with the slow progress" of the federal bill and proposed its own state bill. Morrison noted that although the group lost out in 1989, its bill was killed by a margin of only three votes. This was "great for the first try," she said. Further, although a Massachusetts testing bill may be eclipsed by the state's budget crisis, Troy Soos of the Alliance for Animals said delay is not bad, since the measure continues to gather support from physicians and scientists and has nine cosponsors—considered a lot in Massachusetts.

With companies like Avon announcing an end to testing, things can only get better, Foos felt. In speculating about what would happen if the Maryland bill passed, Branchini reflected, "In a way, we've already won, since companies have been announcing the end of their testing." The uproar over proposed legislation has been valuable in itself, activists seemed to conclude, since it has focused public attention on how products are tested, and may have galvanized companies to work still harder on their research into alternative tests that do not use animals.

III. ANIMAL TESTS FOR CONSUMER PRODUCT SAFETY AND THEIR ALTERNATIVES

In 1933, soon after a woman used the mascara Lash Lure to darken her eyelashes, her eyes began to burn, she lost her vision, her eyes ulcerated, she went blind and eventually died. To prevent such occurrences since, toxicologists at consumer product companies have routinely put cosmetics and other consumer products into the clear, pink eyes of thousands of albino rabbits and observed them for two weeks, and then killed them at the end of the tests. To project other possible dangers to humans, dogs and rats have been fed products until 50 percent of the test group dies, as another measure of product toxicity; other tests for potential dangers have been conducted on animals as well. Today large numbers of animals are still used for product testing, but scientists have devised new alternative tests that are said to address consumer safety concerns without injuring and killing animals in the process.

This chapter traces the history of consumer product safety tests that use animals and the range of alternatives to these tests that companies have started to use.

Historical Background

Although experimenters used animals as early as the 17th century to learn about basic physiology, large-scale animal testing to determine the possible dangers to human health posed by chemical substances did not become common until this century. At first, testing was done for medical purposes, to determine the safe dosages of vaccines and drugs. Consumer product testing gradually entered the picture, but the number of animals used for medical-related testing still far surpasses those used for testing cosmetics and household products.

Although statistics for animal use patterns in the United States do not stretch back to the first days of animal testing, numbers are available from the United Kingdom. Even though the U.K. figures are probably far below the volume of animal use in the United States, it is likely that animal use expanded in both countries at about the same rate, according to Andrew Rowan of the Center for Animals and Public Policy at Tufts University. In 1921, about 20,000 animals were used to test vaccines and biological therapeutics. During the 1920s, the number of animals used went up sharply, Rowan reported in his 1984 book, *Of Mice, Models and Men*, but they still were used mostly for "bioassays of insulin, other hormones and vaccines." By the late 1930s, 365,000 animals were used annually. During the 1950s, the proportion of testing to bioassays was growing, and by 1975, 3.5 million animals were used each year in Britain, many for testing new drugs. Estimates for total animal use in the United States have ranged from 20 million to at least 70 million, but many scientists are skeptical of the higher figure used by some activists. (See Chapter V for more discussion of U.S. animal use estimates.)

Following the 1933 mascara incident, the Food and Drug Administration's mandate to protect public safety was expanded to cover cosmetics, with the Federal Food, Drug and Cosmetic Act of 1938, although specific tests were not required. (See Chapter II for more details.) To ensure that cosmetics were not "adulterated" and therefore illegal, companies turned to contemporary toxicological practices and began to test their products on animals.

Scientists had begun to study eye irritation using animals in the 1920s. After soldiers were exposed to poisonous gas during World War I, scientists studied the effects of mustard gas in rabbit eyes to compare with human experience during the war. In 1938, other experimenters used rabbits to look at more chemicals to determine eye irritation. Finally, in 1944, Food and Drug Administration scientist John Draize standardized the rabbit model and came up with an irritation scoring system, which since has been used in the much-maligned Draize eye test.

At about the same time that rabbits began to be used as models for human eye irritation, the LD50 (lethal dose for 50 percent) got its start. In 1927, a British pharmacologist, J.W. Trevan, was faced with the task of standardizing the production of highly toxic drugs. To come up with a statistically precise measurement of toxicity, he injected groups of rats and mice with different amounts of each drug to see which dose would kill half a test group. Because the drugs were so toxic, precise measurements were necessary, and more than 100 animals were used for testing each new batch of drugs.

Dr. Gerhard Zbinden of the Universitat Zurich has noted that when chemicals began to be widely used in all aspects and products of industrialized society, the LD50 soon became the standard, despite its initial limited application. Although Zbinden says it is "not really understandable why the LD50...became the preferred, and soon the only, procedure accepted by regulatory agencies for the assessment of acute toxicity," his discussion of regulatory classifications of toxicity seems to point to at least one explanation. He says the numerical character of the test provides a convenient way to classify chemicals, even if this number has dubious derivation. "Regulatory agencies fear that if a more scientific, but also more ambiguous, general assessment of a chemical's hazard were used as a classification criterion, the legal basis would become uncertain, and the debate with industry over classification and reclassification of chemicals and consumer products would lead to a never-ending struggle," Zbinden argues. That different countries use different LD50 values for similar hazard categories only proves the regulatory convenience and lack of real scientific merit, he concludes.

During the 1970s, with still more public concern about chemical use aroused by environmentalists and a subsequent spate of federal legislation, product testing accelerated again when attempts to regulate hazardous substances were built into federal regulations. Rowan noted in his book that a major U.S. supplier of laboratory animals, Charles River Breeding Laboratories, in 1981 attributed its sales figures' increase to the new environmental safety laws.

Consumer advocates pushing for safe products helped to initiate expanded testing programs as well, as they raised the stakes in product liability cases. Stephen McNamara of the Cosmetic, Toiletry and Fragrance Association commented on this point at a conference sponsored by the Johns Hopkins Center for Alternatives to Animal Testing in 1981:

> Cosmetic manufacturers, like all consumer product manufacturers in this country, are facing a rising tide [of] litigation in which consumers, sometimes seeking astronomical sums as "damages," allege that they have been injured by defective products. The threat of such litigation is a powerful force impelling manufacturers to test their products before distribution.

> It is important to note that in such product liability actions, evidence of a violation of a health or safety statute or regulation (such as an FDA regulation) may be used as evidence of negligence against a company. As one commentator has observed, "once violation [of an FDA regulation] is established, little more than showing that the product caused the injury may be necessary to establish liability on the part of the manufacturer." Thus, failure to comply with the safety substantiation standards in FDA's regulations and policy statements can "come home to roost" for a cosmetic manufacturer in more than one way. Not only may FDA pursue regulatory action, but an injured customer might cite an FDA regulation in support of a private damage claim.

MILESTONES RELATED TO ANIMAL TESTING

1906 Congress passed the Federal Food and Drug Act, but did not require that products be safe or effective.

1916 Scientists first caused tumors in laboratory animals by feeding them coal tars in a precursor of modern carcinogenicity tests.

1933 One U.S. woman died and many others were injured by Lash Lure mascara.

1937 More than 100 people died from diethylene glycol accidentally used in a sulfanilamide elixir.

1938 Congress passed the Federal Food, Drug and Cosmetic Act, requiring "unadulterated" cosmetics and premarket approval of drugs.

1947 Congress passed the Federal Insecticide, Fungicide and Rodenticide Act, requiring registration of all pesticides.

1959 Russell and Burch published *The Principles of Humane Experimental Technique* in London, defining the concept of alternatives as the three Rs: refinement, reduction and replacement of animal tests.

1960 Congress passed the Federal Hazardous Substances Act to safeguard consumers from hazardous materials, using animal toxicity standards in its definitions.

1961 The drug thalidomide caused birth defects in Europe.

1962 Congress passed amendments to the Food, Drug and Cosmetic Act that required more testing to determine drug safety before marketing. Antivivisection groups in the United Kingdom created the Lawson Tait Trust "to encourage and support researchers who were not using any animals in their research."

1965 The status of alternatives to animal testing was investigated by the parliamentary Littlewood Committee in the United Kingdom.

1967 Activists established United Action for Animals in the United States to promote alternatives to animals, particularly replacement techniques.

1969 The Fund for the Replacement of Animals in Medical Experiments formed in the United Kingdom to promote research into alternatives, and asked scientific organizations to help.

1971 The Council of Europe passed a resolution calling for a center on alternatives to animal use, the "first major establishment breakthrough," according to Andrew N. Rowan.

1972 Congress passed amendments to the Federal Insecticide, Fungicide and Rodenticide Act, requiring a review of all pesticides on the market.

1975 Australian philosopher Peter Singer published *Animal Liberation*, articulating the ideas of an infant animal rights movement.

1976 Congress passed the Toxic Substances Control Act, authorizing the Environmental Protection Agency to require testing for all new chemical substances.

1979 Members of the biomedical research community formed the National Association for Biomedical Research to combat the growing animal rights movement and to promote the idea that animal use is needed for research and testing.

1980 Henry Spira led a coalition boycotting Revlon over its use of the Draize eye test for cosmetics, prompting the company to give $750,000 to Rockefeller University for research into alternatives.

1981 Following agitation from animal rights groups, the Cosmetic, Toiletry and Fragrance Association provided $1 million to create the Johns Hopkins Center for Alternatives to Animal Testing.

1985 PETA picketed Noxell, Avon and International Playtex and criticized animal testing at corporate annual meetings.

1986 The animal rights group Ark II launched a boycott of Gillette, alleging animal abuse in the company's testing program.

1987 IBM, Greyhound and Procter & Gamble received the first shareholder resolutions on animal testing.

1988 The Cosmetic, Toiletry and Fragrance Association started funding a program at Battelle Memorial Institute to evaluate existing alternatives to the Draize eye test.

Congress passed amendments to the Federal Insecticide, Fungicide and Rodenticide Act, requiring the testing of all pesticides on the market within nine to 14 years.

1989 Avon and Revlon became the first major U.S. firms to announce a permanent end to all animal testing of their products. Other large companies also announced animal use moratoriums.

The Johns Hopkins Center for Alternatives to Animal Testing and the Tufts Center for Animals and Public Policy continued efforts to better define how alternatives can be validated. Meetings occured between industry, government and some animal protection activists.

Chemical industry legally challenged government-required animal tests and said other evidence should be considered for cancer risk assessments.

Sources: Andrew N. Rowan, *Of Mice, Models, and Men* (State University of New York Press, 1984), and "Alternatives: Interaction Between Science and Animal Welfare," *Alternative Methods in Toxicology*, Vol. 1 (Mary Ann Liebert, 1983), Johns Hopkins Center for Alternatives to Animal Testing and the Cosmetic, Toiletry and Fragrance Association.

Public insistence on safe products collides with public outrage over how animals are used in testing, however, producing a commonly noted paradox: "Ironically, the public's increasing concern for safety could lead to more testing. Yet it also provides an incentive to develop new techniques, particularly those that promise to be cheaper and faster than current whole-animal methods. A further irony is that developing alternatives, as well as validating them, sometimes requires animal use," wrote the congressional Office of Technology Assessment in its 1986 study, *Alternatives to Animal Use in Research, Testing and Education.*

As the United States began to regulate its products with more extensive consumer safety criteria, other countries followed suit. As a result, one impetus today for U.S. companies to test their products comes not only from U.S. law, but also from foreign standards, some of which are more stringent than those in the United States. A lack of international standardization means that companies usually use the strictest requirements for product testing if they intend to market their goods internationally.

Despite the lack of consensus on testing requirements around the world, there are some signs that a common perspective on animal testing is emerging in some countries. In a recent look at the testing approaches of nations in the Organization for Economic Cooperation and Development, Zbinden concluded that "many of the OECD countries have passed laws regulating animal experimentation in the past 10 years and...a common feature of many of these new laws is the requirement to demonstrate the need for the animal research and testing." A directive on animal use issued in the mid-1980s from the European Economic Community states:

> When an experiment has to be performed, the choice of species shall be carefully considered and, where necessary, explained to the authority. In a choice between experiments, those which use the minimum number of animals, involve animals with the lowest degree of neurophysiological sensitivity, cause the least pain, suffering, distress or lasting harm, and which are most likely to provide satisfactory results, shall be selected.

By the beginning of the 1980s, the animal protection community in the United States had started its frontal assault on the use of animals in product testing. Although attention was given to animals used in biomedical research at first, activists shortly regrouped to concentrate their efforts on consumer product testing instead. (See Chapter I for more on the growth of the animal protection movement and its tactics.)

The pressures placed on companies by well-organized activists, plus the knowledge that non-whole-animal tests probably would be ultimately cheaper and perhaps better, set the stage for a decade of research and development for alternative product safety tests. Further, in vitro toxicology has flourished in part because of advances in tissue culture techniques and other analytic methods that can measure tiny amounts of biological material. The concept of alternatives has been refined in the last 10 years, as well, although the basic idea of alternatives has been around for at least 30 years. A new scientific, governmental and corporate perspective on animal testing has emerged, and many authorities now have shifted from a defense of existing testing methods to open support for alternatives to the use of live animals.

Research on Alternatives

Funding sources for research on alternatives have run the gamut from anti-vivisectionists to industry groups. An important research effort in the United Kingdom got underway long before any such work was done in the United States. The Fund for the Replacement of Animals in Medical Experiments was founded in 1969. It is described in a 1983 book published by the Academic Press in London, *Animals and Alternatives in Toxicity Testing*, as an "independent charitable trust" that "seeks to avoid confrontation" and works to find alternatives to animal use. FRAME formed a committee of experts in 1978 to review and assess current toxicity tests. The committee's recommendations were published in a comprehensive report in 1983. FRAME's quarterly newsletter, *ATLA*, serves as a major source of information on the status of alternatives research and implementation.

The first major U.S. funding for alternatives came about 10 years after FRAME's founding, prompted by the highly visible campaign against use of the Draize test at Revlon in 1980. The company contributed $750,000 to Rockefeller University for research into alternatives. Soon after, the Cosmetic, Toiletry and Fragrance Association provided $1 million to establish the Johns Hopkins Center for Alternatives to Animal Testing, which has grown to be an important clearinghouse for alternatives work in the United States and abroad. CAAT's annual conferences are compiled in the only book series devoted to alternatives in toxicology, published by Mary Ann Liebert in New York. By the end of the 1980s, CAAT had provided seed money grants of up to $20,000 to more than 100 research projects, researchers funded by CAAT had published more than 100 book chapters or articles, the Center's newsletter subscription base was 22,000 worldwide, more than 90 presentations on the subject of alternatives had been made around the world, more than 76 companies had contributed more than $4 million to the Center, and more than a dozen animal protection groups had made contributions to the Center. In 1989, the Center received its first federal funding with a $350,000, five-year grant from the National Institutes of Health, which will be used to support its core activities.

Other funding from animal advocates in the United States came earlier, when the New England Anti-Vivisection Society gave $100,000 to a researcher at Tufts. The American Fund for Alternatives to Animal Research has donated about $250,000 to research since 1977, and the Millenium Guild of New York set up two $250,000 prizes to award research on an alternative to the Draize Test and to fund alternatives research in general, although only about $10,000 of the Millenium Guild's money has been disbursed so far. Another research effort funded by animal protection activists is the International Foundation for Ethical Research, which is supported by the National Anti-Vivisection Society and the American Anti-Vivisection Society.

The Cosmetic, Toiletry and Fragrance Association announced in December 1987 that it would start its own program at Battelle Memorial Institute in Columbus, Ohio, to validate 14 existing nonanimal tests that might be substituted for the Draize eye test. An industry report on the move said, "The new program marks a shift from CTFA's current emphasis on basic research in animal alternative testing." CTFA's full financial support for the Johns Hopkins CAAT continued until October 1988, when the industry association shifted funds to its own effort, which it said would identify nonanimal testing procedures for specific product lines. The CTFA still gives money to CAAT, but at reduced levels, and individual companies fund the Center directly.

The Soap and Detergent Association is funding another industry effort to validate alternatives to the Draize eye test for specific product categories made by its member companies. By the end of 1989, the SDA's program had gone through two of three projected phases. Phase I evaluated 14 proposed tests by comparing their performance to animal data for eight test materials. Six tests passed the initial stage and were sent on to Phase II, where 15 test materials were used. The program started its third phase in June 1989, also using six in vitro methods, and expanding the number of test materials to include 40 more substances. Results from Phase III are expected in summer 1990.

In addition to the efforts of the CTFA and the SDA, individual companies are conducting in-house research on a variety of nonanimal safety tests. Procter & Gamble has been a leader in the search for alternatives, and spent $3.5 million in 1988 alone on its in vitro research programs. Avon conducted research for nine years before deciding to use the Eytex method as a replacement for the Draize eye test, in conjunction with an extensive computer database. Colgate-Palmolive helped to develop a test using chicken egg membranes instead of rabbit eyes. (For more on overall and specific company efforts, see Chapter IV.)

Government efforts examining nonanimal tests have been complicated because agency mandates differ widely. In the last few years, however, an interagency group with representatives from the Food and Drug Administration, the Consumer Product Safety Commission and the Environmental Protection Agency have met periodically and informally to assess the status of alternatives and to determine what kind of regulatory approach to take. Also, a fall 1988 workshop on alternatives to the Draize eye irritancy test coordinated by the Soap and Detergent Association was jointly sponsored by these three agencies and three other industry groups: the Cosmetic, Toiletry and Fragrance Association, the Chemical Manufacturers Association and the Chemical Specialties Manufacturers Association. Participants in the workshop hailed it as a ground-breaking meeting between industry and government that showed promise for further cooperation on alternatives work.

In 1989, the Environmental Protection Agency gave a two year grant of $270,000 to the Alternatives Project at the Center for Animals and Public Policy at Tufts University. The Alternatives Project is a two pronged effort. With additional assistance from Bristol-Myers Squibb and Tom's of Maine, it is publishing a bimonthly newsletter, *The Alternatives Report*. In addition, a series of workshops is underway with representatives from industry, government regulatory agencies and animal protection groups, to develop a consensus for the validation of alternatives for safety testing. Center Director Rowan told IRRC that he foresees a symposium at the end of the grant period when a core group will present policy recommendations formulated from information gathered at the workshops. He noted that many people are unhappy with the intense polarization of this issue, and that he hopes the project will bring groups together to work constructively.

Animal Testing Versus Its Alternatives

Although nearly everyone party to the debate over animal testing can probably agree that there are serious flaws to the practice, it does have some advantages. The head of the Johns Hopkins Center for Alternatives to Animal Testing, Alan Goldberg, points out "the most obvious and important advantage of whole-animal testing: It provides an integrated biological system that serves as a surrogate for the complexities of human and other animal systems." Scientists can observe how the entire organism responds to a substance, including its behavior, the interaction of its cells and tissues, the growth and recovery of damaged tissue and how the substance is excreted by the body. Another advantage is the extensive historical database of animal toxicity information that covers a wide range of substances and stretches back as much as 50 years; these data provide a convenient reference to evaluate the relative dangers of different chemicals. Finally, despite the growing acceptance of alternative methods, many toxicologists—even those who are ardent supporters of nonanimal methods—still believe that animal tests remain the best available way to assess possible risks to human health.

The main disadvantages to animal tests are listed in an August 1989 article in *Scientific American* by Goldberg and his colleague at CAAT, Assistant Director John Frazier, as follows: "Animal discomfort and death, species-extrapolation problems and excessive time and expense." Animal protection advocates stress animal pain, saying that many laboratories submit animals to inhumane treatment without anesthesia or adequate consideration for their suffering. They also argue that tests use an excessive and unnecessary number of animals in a kind of "overkill" to pass toxicological guidelines. Extrapolating the toxic effects of substances from one species to another can be done in some instances because there are physiological similarities in the way chemicals react between species, but such predictions nevertheless are not always certain. Purchase and maintenance costs of laboratory animals are high, and, the Office of Technology Assessment noted, "animal methods generally are more labor-intensive and time-consuming than nonanimal methods, due to the need, for example, to observe animals for toxic effects over lifetimes or generations. Testing can cause delays in marketing new products, including drugs or pesticides, and thus defer a company's revenues." A further problem is the lack of standardization of animal techniques.

In their *Scientific American* article, Goldberg and Frazier also lay out the pros and cons of in vitro techniques, one major category of alternatives. Aside from the obvious escape from animal use, these are: 1) in vitro tests are easier to standardize, 2) it is "easier to establish the critical concentrations of toxins" and 3) "novel compounds available in limited amounts can be tested, and disposal problems are minimized if a compound turns out to be toxic." Further, Goldberg suggested to IRRC that "maybe the only way to evaluate the toxicity, and in some cases efficacy, of [the products of advanced bioengineering] will be in cultures of human cells." Disadvantages to nonanimal methods are that it is very difficult to mimic an integrated living system, and the range of historical data available for animal tests does not yet exist for in vitro methods and computer analyses. Although these are serious drawbacks, promoters of the new techniques say they can be overcome; other analysts counter that predictions of complete animal replacement in the next few years are simply wishful thinking. What is clear is that computer and in vitro approaches to product testing can reduce the numbers of animals used and refine the tests.

Dr. Peter Medawar's statement from 1969 provides one reflective assessment of the debate:

> The use of animals in laboratories to enlarge our understanding of nature is part of a far wider exploratory process, and one cannot assay its value in isolation—as if it were an activity which, if prohibited, would deprive us only of the material benefits that grow directly out of its own use. Any such

prohibition of learning or confinement of the understanding would have widespread and damaging consequences; but this does not imply that we are forevermore, and in increasing numbers, to enlist animals in the scientific service of man. I think that the use of experimental animals on the present scale is a temporary episode in biological and medical history, and that its peak will be reached in 10 years time, or perhaps even sooner. In the meantime, we must grapple with the paradox that nothing but research on animals will provide us with the knowledge that will make it possible for us, one day, to dispense with the use of them altogether.

One factor promoting the development of alternatives has little to do with ethical questions about animal use, although it is mentioned by activists in their arguments for nonanimal tests. Many more chemicals exist and are being created than have been tested using traditional animal methods. There is little likelihood that they can be tested using current techniques, since the time, laboratory personnel and money do not exist to make safety assessments for all the synthetic chemicals now in circulation.

Rowan noted that between 1940 and 1977, "annual production of synthetic chemicals rose 300-fold to 306 billion pounds." One estimate in the mid-1980s found that no toxicity information was available for 56 percent of cosmetic ingredients, 78 percent of chemicals in commerce that account for at least 1 million tons of production annually, 38 percent of pesticides and pesticide ingredients and even 25 percent of drugs and materials used in drugs. Current animal toxicity studies cost from $500,000 to $1 million per chemical, take two to three years to complete, and use thousands of animals, Goldberg and Frazier say. The 1986 OTA report concluded, "Not all [chemicals] must be tested, but toxicologists must expand their knowledge of toxic properties of commercial chemicals if human health is to be protected to the extent the public desires." Clearly, faster techniques are sorely needed; alternative toxicity tests can provide a happy solution, according to a growing number of toxicologists.

The Three Rs and Alternative Tests

While most of the progress away from traditional animal toxicity tests has come in the last decade, the desirable approach to alternatives was defined at least by 1959. In *The Principles of Humane Experimental Technique*, W. Russell and R. Burch categorized alternatives by what are now known as the "three Rs": reducing the number of animals in tests, refining tests to decrease pain and replacing methods that use animals. Although the pigeon holes of the three Rs do provide a convenient and broadly accepted approach for assessing the range of possible alternatives to animal tests, their classification system is not as neat as it seems. The Rs overlap and cannot always be clearly distinguished as separate conceptual entities.

Animal testing critics do not weigh all three Rs equally, either. Although animal welfare advocates are more likely to be satisfied with an incremental approach to implementing alternatives, animal rights proponents hold the more purist view that no testing is acceptable, so variations on tests (refinements and reductions) do not produce better tests, because all tests are equally bad. In their view, replacements are the only acceptable alternatives. This perspective makes most company officials who have tried to reduce and refine their animal use apoplectic. A June 1989 letter from the Cosmetic, Toiletry and Fragrance Association illustrates the position of many companies: "We are not dealing with rational opponents. We are dealing with zealots...who distort our industry's animal testing methods; who duplicitously misstate FDA requirements; who smear the industry and companies trying to do their best to find alternatives to animal testing." Despite deep-seated corporate frustration with the most vocal critics, companies have gone ahead with efforts to change their animal use patterns, and many have made clear progress in reducing and refining their tests. In a few cases, companies have combined some of the available alternative techniques and have replaced their animal use completely.

Although the three Rs serve as a common reference point in discussions about alternatives, the range of proposed alternatives is broad. One general suggestion from FRAME is to examine carefully why tests need to be conducted at all, and to make sure that if animals are used, they are properly cared for in the laboratory. Modifications of traditional animal studies, including statistical techniques that reduce the number of animals in tests, fall under the alternatives umbrella, according to some observers. Other methods, which do not use whole animals, are in vitro tests, computer-generated predictions of biological responses to chemical substances, databases to avoid unnecessary repetition of tests, and physical/chemical tests of substances. These "replacement" methods are also used to reduce the numbers of animals used and refine the kinds of tests that are done on potential products. The use of less sentient organisms and human studies are also included in some lists of alternative techniques. To provide the broadest possible look at what is meant by the term "alternatives," the listing below is intended to cover the range of proposed techniques that are used in some way to modify traditional whole-animal testing methods.

Modifications and refinements of traditional animal studies: Statistical methods modifying traditional tests such as the LD50 are responsible for reductions in animal use of up to 90 percent, with only a dozen animals or fewer used per substance, in contrast to traditional techniques where more than 100 animals were used. FRAME stresses the early and extensive use of statistical techniques to improve experimental design and reduce animal use. Also, methods to decrease the pain and distress felt by animals in experiments are becoming more common with the improved use of pain relievers. Other modifications include the enhanced diagnosis of pain or internal toxic reactions such as tumors or organ deterioration by using modern medical technology, using lower doses of test materials and timely euthanasia. Further, as scientists refine their use of species that are most predictive of human reactions to a given substance, they tend to use fewer animals.

In vitro techniques: Goldberg and Frazier say that although "in vitro [literally] means 'in glass'...biologists interpret the term more broadly to mean research that does not involve intact higher animals." Thus, methods ranging from cell and organ cultures to physicochemical tests to bacterial assays fall under this classification.

Cell, tissue and organ cultures are being developed for in vitro tests that measure reactions to toxins and would substitute for whole animal reactions. Although culture techniques do require living tissue that must come from either animals or humans, one animal can be used as the source for a number of different tests, and human tissue left over from surgery or cadavers also can be used. The possibility of using human tissue makes in vitro methods attractive to toxicologists because although these systems are much simpler than whole animals, one major difficulty in toxicity tests up to this point has been the extrapolation of animal reactions to humans. OTA noted that some toxic reactions in vitro are measured by how many cells exfoliate from a tissue's surface, by counting the number of cells and nuclei after they are exposed to a substance, and by monitoring cell membrane integrity by seeing if the cells take in nutrients; many different endpoints exist, however.

Problems with cultures are that some cells cannot be cultured yet, some cells change their character when they are cultured, and the supply of human cells so far is limited. FRAME recommends the establishment of tissue banks that would provide standard tissue types so that in vitro culture methods can be standardized; such banks would also help to alleviate the supply problem. OTA noted in its 1986 report that tissue cultures are more difficult to use than cell cultures, because they are harder to sustain; recent progress in in vitro technology has surmounted some of these difficulties, however. In a 1989 paper, Goldberg concluded, "The use of culture methods must be developed and implemented cautiously," but "culture techniques have great potential."

Despite the ultimate replacement potential of in vitro tests, in some cases animal use can increase while companies search for these alternatives, however, because the new studies are evaluated beside old methods for comparison and validation. Avon, for instance, disclosed that 43 percent of the mice it used in 1985 were for in vitro experiments and that overall use of mice grew 12 percent the same year (compared with a reduction that year in its use of other animals). It is expected that scientists eventually will be able to maintain various self-sustaining cell cultures without reliance on whole animals, though.

In vitro methods have the potential to replace some animal tests, but they are also used in conjunction with in vivo (whole animal) tests as screens. Because each in vitro system measures just one kind of reaction out of a broad array of possible reactions in whole animals, it is not realistic to talk of replacing a single animal test with a single in vitro test; rather, a battery of in vitro tests, perhaps used in conjunction with other alternative methods, could theoretically replace a whole animal test. Each in vitro test is just one piece of the whole toxicological puzzle.

Physicochemical methods, although put in a separate category from in vitro tests by some observers, in essence do fall under the in vitro roof. These methods look at specific chemical reactions that may occur to a substance to determine potential toxicity. One example of this method is the Eytex system, from the National Testing Corp. of Palm Springs, Calif. Eytex predicts eye irritation potential by measuring the reaction of an artificial protein matrix to chemicals, the company says. Avon uses the Eytex system, along with its database, to replace the Draize test for eye irritancy, but Procter & Gamble conducted experiments with Eytex and was not satisfied with its performance. (See below for more on Eytex.)

Goldberg says there are three basic approaches to in vitro methodology Empirical studies try to establish correlations between in vitro and in vivo mechanisms. He notes that "results tend to be somewhat unpredictable" because there is no understanding of what makes the test substance create adverse effects. Model systems try to mimic living systems, and although they usually are "neither complete nor faithful in all aspects of that system being modeled...[they] tend to provide useful information if the data are not overinterpreted," Goldberg says. Mechanistic studies try to establish why toxic reactions occur by looking at how they work in living systems. If all the possible ways chemicals can react with biological tissue are identified, and individual in vitro systems are established that will test for each of these toxicity mechanisms, then the resulting menu of tests can be used to predict toxicity. This is no easy task, though. Animal tests are mainly empirical, not mechanistic, and so far the mechanisms of toxicity have not been fully understood; therefore, in vitro tests promise to add significantly to scientific understanding of toxicity.

Further, now that in vitro toxicology is becoming a more concrete reality, a growing number of small firms are marketing products for in vitro toxicity studies and signing agreements with large companies, who are using the in vitro-related products in testing programs. Arthur Benvenuto, the chief executive officer of Marrow-Tech, in Elmsford, N.Y., is quoted in a March 1989 issue of *Business Week* as saying that products such as his have a potential $1 billion market. Marrow-Tech is marketing its skin cultures for use in irritancy tests and announced a collaborative agreement with Colgate-Palmolive in June 1989. Clonetics, in San Diego, provides its clients with packs of human cells for use in vitro. It recently introduced the Neutral Red Bioassay, which uses one of the company's cell packs, EpiPack, to diagnose cellular toxicity. A Massachusetts company, Organogenesis, in Cambridge, has signed agreements with Avon, Estee Lauder and Mary Kay Cosmetics for "collaborative research and supply agreements" in conjunction with its Testskin product, a patented skin culture system that is used for product testing and research. Another Massachusetts company, Millipore, in Bedford, sells membranes in plastic wells to provide the environment for cell cultures for toxicity testing. A company representative told IRRC that several major companies are customers, but he would not name them.

Computers: *Alternative News,* a U.K. newsletter supporting alternatives to animal testing, gave the following explanation of computer models:

> To set up a computer simulation, all the available results on a particular substance, or series of related substances, are extracted from...scientific publications. Equations can be made which link the structure of chemicals...with their effect in the body. The link between molecular structure and effect is called the structure-activity relationship, and the computer model, using the structures of known substances, predicts what the effects of related, novel chemicals might be.

FRAME says there are basically two kinds of computer models. In the first, "Changes in physical properties...are correlated with biological response" for a "training set" of chemicals. A numerical relationship is then defined for use with chemicals similar to the training set. In the second sort of model, toxicity potential is figured from the actual structure of chemicals. OTA explains that "frequently used properties [in computer models] include an affinity for fats versus water,...the presence of certain reactive groups, the size and shape of molecules, and the way reactive fragments are linked together." The structure-activity relationship (SAR) can predict the probable biological response, help experimenters decide if a set of chemicals is worth developing further, prioritize substances for testing, verify in vitro testing, and provide other information about handling the test substance. The ultimate challenge for SARs is to mimic entire living organisms fully so that toxicity potential can be fully evaluated.

The use of computers for storing and providing easy access to existing toxicity information should not be underestimated, many experts say. Toxicologists commonly conduct literature searches to find information about the chemicals they are studying and use databases to infer information about these chemicals, thereby avoiding unneeded test repetition. Many companies maintain in-house databases to store information about the ingredients and formulations used in their products for future reference. This practice has important implications for reducing animal use. Avon, for example, says its extensive computer system is the mainstay of its move to nonanimal safety assessments. Also, the Cosmetic, Toiletry and Fragrance Association has been building a database of information on cosmetic ingredients called the Cosmetic Ingredient Review, compiled from information voluntarily supplied by its member companies.

Kurt Enslein, president of Health Design Inc., a small Rochester, N.Y., company that is marketing a computer software package called Topkat for predicting toxicity data, says SARs predict toxicity correctly from 85 to 97 percent of the time. HDI notes that its Topkat program is "heavily dependent on the quantity and integrity of available data....As a result, the models developed by HDI reflect the quality of research which has been done with chemical structures." However, Topkat can be used to replace animal use for the LD50 and Draize eye and skin tests; it can also predict carcinogenicity, mutagenicity and teratogenicity, among numerous other applications, the company says.

Human studies: FRAME suggests that human studies are an important component that cannot be overlooked. It recommends controlled exposure of substances to humans to determine potential toxicity, as well as extensive post-marketing surveillance.

Less sentient organisms: Less sentient organisms, which have less developed pain sensitivities, can be used for testing chemical compounds. For example, horseshoe crab blood is now used to test whether drugs cause fever as a side effect. Microorganisms can also be used; alterations of protazoan behavior after the application of a test substance can reveal potential toxicity. One test that follows this approach is the tetrahymena inhibition assay (described below).

Combinations of alternatives seen as best approach: Experts suggest that a tiered approach to substance testing is a good way to use existing alternative methods. Goldberg has described one such approach: The first step could be a computer-aided literature search to ensure that tests had not been done on the substance already, as well as to look for available information on chemicals similar to the substance. Next, a battery of in vitro tests could be run with the substance to determine whether further development was advisable. Depending on the intended use, the chemical might be considered too toxic for further development, so extensive testing could be avoided at this relatively early state. Finally, if enough information had not been gained by the first two steps, a final in vivo test could be used.

In another strategy, Diane Benford of the University of Surrey noted in a FRAME report that combinations of in vitro and in vivo techniques can be used to take into account factors that do not show up in vitro, such as metabolic changes and other ways in which chemicals interact with the body. A substance can be administered to an animal, then some of its tissue can be "removed, cultured and examined, e.g., for DNA repair," she says. The end result would be fewer animals used and less suffering for animals that are employed. Despite the difficulties with simulating the complexity of living creatures, Goldberg told IRRC: "Whether five, 10 or 15 years from now, my guess is there will be a time when animals used in product safety will be few or none at all."

Animal Tests: Criticisms and Alternatives

The two tests most vilified by animal protectionists are the LD50 and Draize, but other tests are used to make safety assessments for consumer products as well. The main reason why product tests are conducted is to assess the possible dangers to human health--to assuage concerns about risks to public safety and to avoid product liability suits. The two factors in the risk assessment equation are how dangerous a substance is, an estimate produced by testing, and what the likely exposure is, which depends on intended product use and possible misuse.

The kinds of tests done depend on the product. When a company is developing a product, its toxicologists decide what kinds of tests to use by looking at the probable routes of exposure to people, to determine how the product would affect the user. "People don't put paint stripper around their eyes and wear it around all day," so the same kinds of concerns that the Food and Drug Administration has for cosmetics do not exist in that case, an official from the Consumer Product Safety Commission told IRRC. Edward Jackson of Noxell Corp. explained, "Cosmetic products are applied to discreet body parts, such as fingernails, as well as to the entire body, such as bath preparations. With manicuring preparations, frequent hand to face contact means product exposure to the facial skin and even the eye. With bath preparations, we must consider exposure of the product to the entire surface of the largest organ of the human body, the skin, and even the mucous membrane lining of certain cavities." The table below shows the typical range of tests that might be conducted for cosmetics, as an example.

Common industrial safety testing practices
for the cosmetic industry

Review of ingredients	*Tests for products*
1. Published studies on the known effects produced by the ingredients are examined.	1. Animal toxicity and irritancy tests are conducted of neat chemicals and chemical classes.
2. Scientific literature reviews and ingredient safety monographs of the Cosmetic Ingredient Review program (from CTFA) are studied.	2. Human patch tests are done.
	3. Controlled human use studies are conducted.
3. Company data on ingredients, fragrances and colors are reviewed.	4. Regional marketing surveillance takes place.
	5. National marketing surveillance occurs.

Source: After Edward M. Jackson, "Alternative Practices in Safety Testing," *Alternative Methods in Toxicology Vol. 1* (Mary Ann Liebert, 1983), p. 63.

Other types of products raise different problems. Liquid household cleaner could be splashed in the eyes or on the skin, or accidentally drunk by an errant toddler; the company therefore would want to know how human eyes and skin could be affected, possible allergic reactions and the probable effects of a relatively large dose of the product if it were ingested. If people are likely to be exposed to a product often over a long period, the company also might want to know the possible chronic effects: Could it affect human fertility, cause birth defects or cancer or adversely affect the body in some other way? A typical battery of animal tests includes acute toxicity tests, which are conducted mainly to determine the effects of massive accidental exposure, and some long term studies.

Short term toxicity tests: The following section describes the methods used for product testing and the specific criticisms of these tests and provides examples of currently available and proposed alternatives.

The LD50—In the classical version of this acute toxicity test, up to 100 or more animals in each test are exposed to the test substance by oral ingestion or inhalation, or in rare cases by dermal application. Rats are used most often for oral ingestion studies, but rabbits and sometimes guinea pigs are used in acute dermal toxicity tests. After scientists make initial estimates of the toxicity range of the test substance, different doses are tried on groups of animals until 50 percent of one group dies. The dose this group received is the precise median lethal dose, or LD50. The Humane Society of the United States reports that oral ingestion is the most common method. HSUS says this method "produces signs of poisoning including bleeding from the eyes, nose or mouth; labored breathing; convulsions; tremors; paralysis; and coma." Scientists dissect the animals that die to determine the cause of death. After two weeks, the surviving animals are killed and examined, as well, so that scientists can learn how the substance affected their bodies.

In acute toxicity tests, "the investigator is observing an immediate response in which the organism's defense mechanisms are rapidly overwhelmed....One of the functions of acute testing is the identification of unexpected toxic effects. The empiricism of this approach requires that a relatively good model for the whole human being be used. This generally means using a whole mammal," Goldberg said in a recent paper. He does not, however, advocate the use of the classical LD50 technique with large numbers of animals, and he notes that "the classical LD50 test (with a few specific exceptions) appears to be on its way out as a regulatory requirement" because of scientific criticism and pressures from the animal protection community.

Because precise LD50 statistics are the foundation of many countries' regulatory guidelines, large companies marketing their products internationally tend to follow the most stringent guidelines and may conduct classical LD50 tests even if they are not required by domestic U.S. law. As OTA commented, "The biggest obstacle to limiting or eliminating use of the LD50 is institutional: Many regulatory schemes rely on it for classifying substances." Animal testing statistics calculated for Britain for 1987 show that of the 122,000 animals used in subacute and acute toxicity tests, 111,000 were used in classical LD50 procedures. Commenting on these figures in March 1989, FRAME attributed the 50 percent use of the classical LD50 to foreign regulatory requirements for precise lethal dose values. The U.K. statistics emphasize the need for international regulatory standards, observers say.

—*Criticisms:* Critics of the LD50 maintain that there are a number of serious flaws in the test and that it should not be used as a way to predict toxicity in humans. They say:

- The numerical value of the LD50 is not a biological constant, but is highly influenced by factors not related directly to the experiment, such as the sex of the animals, their diet and the laboratory where the test is performed. OTA reports, "The LD50 for two neighboring [regulatory toxicity] levels typically differs by a factor of four to 10. Yet, the reproducibility of test results does not justify even these distinctions."

- The LD50 test concentrates on when animals die and not why they die. Knowledge of the morbidity induced by a substance that can be obtained from clinical human studies is more useful, especially when treating accidental poisoning victims. FRAME says fewer animals can be used with more precision, if careful attention is paid to the information that can be obtained on sublethal effects, and not too much weight is put on the numerical indices produced.

- The LD50 determined in animals does not predict the lethal dose in humans very well because species differ and interspecies extrapolation of toxic effects is highly problematic. Zbinden points out that "data obtained in properly performed acute toxicity studies have a high degree of predictability for many of the functional disturbances and organ lesions occurring in humans" and other animals, but he does not favor the test as a broadly used regulatory tool because of its many other flaws.

- Animals experience extreme pain and distress. Some experimenters say the administration of painkillers would interfere with test results, because all the effects of the substance would not be seen.

- Acute toxicity tests might record deaths from nontoxic causes because substances that are less toxic, such as cosmetics, must be forced into animals in such large quantities that they rupture or block internal organs. Although some critics make this point, Rowan said in *Of Mice, Models and Men* that "stomach rupture is an uncommon form of death in LD50 testing," and that it occurs because of improper administration of the test material rather than the amount of material.

—*Alternatives:* Methods that would completely replace acute toxicity studies in animals are difficult to find, because these tests measure how the entire body reacts to a substance, as noted above. However, the classical LD50 has numerous variations that are categorized as alternatives since they have drastically reduced the number of animals used, by as much as 90 percent. Experts say that combinations of in vitro tests and computer information might eventually be used to replace acute toxicity tests completely. This would mean overhauling existing regulatory structures that depend on LD50-style statistics as a measure of toxicity, however. Diane Benford of the University of Surrey wrote in a 1987 *ATLA* article, "Animal LD50 values will never be predicted from cell toxicity," since the LD50 statistic itself is so variable and because the way chemicals work on an integrated living system cannot be completely mimicked in vitro. Thus, she said, "The value of LD50 data is limited and the relevance of LD50 replacement tests is therefore also questionable." The challenge for scientists and regulators is to devise a new way to quantify toxicity levels and to integrate the new approach into regulatory mechanisms.

In the meantime, tests with fewer animals are being used to generate the LD50-style statistics that still remain widely accepted. Most of the LD50 alternatives now in use "increase the dose sequentially, thereby allowing the experiment to stop when a certain limit is reached. Thus, fewer animals die in the conduct of a test, but its duration could increase from two weeks to a month or so," OTA wrote. Today, some progress is being made on alternatives that would assess acute toxicity without using animals. Computer-generated toxicity predictions are more common, and further progress has been made using cell culture techniques.

Alternatives to the LD50 include:

- *The Limit Test:* About five rats or mice are given a single high dose of the test substance. If no animals die, there are no more tests and the substance is considered nontoxic. The general nature of this method is considered appropriate, OTA said, because "if a substance is not lethal at high doses, its precise LD50 is not very important...." By 1985, the number of animals used by cosmetics companies had dropped 75 to 90 percent after firms switched to using this technique, reported *The Rose Sheet*, an industry publication.

- *Approximate Lethal Dose:* Four to 10 animals are typically used in this test. Each animal gets a dose that increases in volume by half again as much as the previous dose, until the toxic dose is found. Use of this technique is also fairly common.

- *Up-Down Test:* About six animals are used. "Each animal receives a single dose...[which] is lowered after signs of severe toxicity develop or is raised after an animal survives one week without such signs. The resulting information is evaluated in a commonly available computer program" and provides an estimated LD50 value, the Humane Society of the United States reported.

- *Computer models:* No animals are needed for these tests because predictions of the toxicity of a chemical can be based on computer analyses of its structure and properties. However, the computer data are initially generated from animal experiments. Health Design Inc. says that its Topkat program "can provide toxicity predictions" for oral and inhalation LD50 values and EC50 (effective dose for 50 percent, or the dose that makes 50 percent of a test population sick) values. The company reports that this information is obtainable from the program by entering into the computer the structural code of the test compound. Several large chemical companies and government agencies, including the Food and Drug Administration and the Environmental Protection Agency, are using Topkat, HDI reports.

- *In vitro methods:* Cell and organ cultures can measure acute toxicity to a certain extent, and can screen out very toxic substances. Animals are needed to supply the initial tissues for cultures, but scientists are refining ways to maintain and propagate cell lines indefinitely. Reproducing organ cultures is a considerably more difficult problem that has yet to be solved, but "many organ cultures usually can be derived from one animal," the Humane Society says. However, the acute toxicity assessment need to measure systemic reactions poses problems. Still, the British Toxicology Society has determined a Fixed-Dose Procedure with nonlethal endpoints. Tissue cultures can measure interactions in systems to a certain extent, according to the Humane Society: "Tissue from one organ can be exposed to specific hormones produced by other organs, or a potentially toxic chemical can be incubated with liver cells to determine whether the liver detoxifies the chemical before it can exert any toxic effects on other cells." Because systemic reactions to compounds are hard to identify, though, in vitro methods so far are used only for screening, in conjunction with in vivo tests, and to understand mechanisms of toxicity.

- *Tier testing*: A combination of the above methods can cut animal use. Alternative methods can be used to look at a substance's effects on cell and tissue cultures in initial testing stages, screening out chemicals that seem to be too toxic for their intended use. A final stage of in vivo testing can pick up other toxic effects that might not show up in the in vitro studies, and refine the toxicity assessment further.

The Draize eye test—Six rabbits are generally used now, although previously as many as 18 could be employed in one test; albino species are preferred because their clear eyes make observation easier. The lower eyelid of one eye on each rabbit is held away from the eye, the test substance is dropped in, and the eye is held shut for a short period so that the substance will be spread around. The other eye is left untouched as a control. Collars or stocks are used to keep the rabbits from touching their eyes. Over the next two weeks, lab workers evaluate redness, swelling, hemorrhage or corrosion, sometimes using a set of color slides to score the irritation. When the test is over, the rabbits are killed or used for another test.

—Criticisms: Animal advocates and scientists alike see problems with the test. They maintain:

- The eyes of rabbits used in the Draize test differ substantially from human eyes because rabbits have a higher threshold of pain and therefore do not react the same way humans would; Bowman's membrane, part of the cornea, is six times thinner in rabbits than in humans; and the rabbit cornea makes up 25 percent of the eye's surface, while the corresponding area in the human eye is only 7 percent. Also, a nictitating membrane, or third eyelid, has "an uncertain effect on a chemical's contact with the eye."

- Test results vary widely, because of different scoring assessments from different laboratory workers, small test populations, and minor changes in test protocols, and because the test responds better to severe irritants than to moderate ones. A comparative study of how 25 laboratories evaluated application of the Draize test's grading system concluded that "the procedure should not be recommended as the standard for any new regulations."

- The test is expensive and time-consuming in terms of personnel, laboratory time and facilities.

- Rabbits experience pain and suffering and are eventually killed.

—Alternatives: Of all the techniques used for consumer product testing, the largest number of possible alternatives seems to have been generated for this method. Draize eye test alternatives are the subject of in vitro assessment programs being conducted by the Cosmetic, Toiletry and Fragrance Association and the Soap and Detergent Association, and companies are starting to replace their ocular irritancy tests completely with nonanimal methods. The SDA program noted in an intermediate assessment of the alternatives it is studying that it did not expect to replace animal use completely in the immediate future, however. Animal use for eye irritancy tests has decreased 87 percent, according to one recent report from Tufts. Many of the proposed alternatives are being used for eye irritancy assessments, but participants in a fall 1989 workshop sponsored by the Tufts Center for Animals and Public Policy "felt that further decreases in the use of test animals are likely to be much smaller and more difficult to achieve."

Alternative methods include:

- *Modifications of animal tests:* The efforts made by Avon before it stopped animal testing completely are typical of procedures that modify the Draize eye test but still use animals. The company told IRRC that animals were held by hand instead of put in stocks, an anesthetic was applied to the eye before the test materials were applied, and test materials likely to be irritating at full strength were diluted. The low volume test was originally developed by Dr. John Griffith of Procter & Gamble. CAAT reported in 1988 that "anecdotal human data from P&G consumer hotline calls suggest the low volume test is more predictive of human response time than the traditional Draize. In other kinds of modifications that still use animals, the Humane Society of the United States says that eye irritation can be gauged without killing the animals after the tests to perform necropsies.

- *Skin irritation testing:* Before materials are tested in the eyes of animals, a skin patch test on either animals or humans can be done. If the substance is classified as a dermal irritant, it often is assumed to be an ocular irritant also, and no further tests are needed.

- *Measurement of pH:* A practice becoming more common is to screen the material before in vivo tests by measuring its pH. If the reading is 2 or less (very acidic) or 12 or greater (very alkaline), ocular irritancy is assumed and no further tests are needed.

- *In vitro methods:* A host of in vitro methods are being developed and evaluated by many investigators. A few examples of these techniques are:

  - *CAM/CAMVA:* Irritation is measured on the membrane of fertilized chicken eggs. In a fall 1989 assessment of the CAM assay, *The Alternatives Report* said, "A number of differing protocols have been developed and utilized in irritancy evaluations. Results from these studies have ranged from dismal to very promising according to the particular investigator." Colgate-Palmolive has done extensive research on this test, which is used by a number of companies to screen for irritancy.

  - *Eytex:* This physicochemical test depends on a mechanistic understanding of how corneal opacity occurs. Researchers "construct[ed] a synthetic matrix of proteins that...behave[d] in the same way as the in vivo cornea," *The Alternatives Report* said in an article on Eytex in summer 1989. When test compounds make the protein matrix opaque, the degree of opacity indicates how severe an irritant the test material may be, Eytex promoters say. Newly developed variations of the test are said to allow for testing of chemicals with several kinds of physical characteristics. Although *The Alternatives Report* noted that Eytex "appears to be winning some important converts and is doing very well in the correlation stakes," Procter & Gamble researchers were not satisfied with the test's predictive ability for their products when they evaluated it.

- *Cell cultures:* A wide range of tests that measure cell toxicity have been developed and are being constantly refined for eye irritation studies. In general, these techniques measure the effect of different dilutions of a test substance on a specific number of cultured cells, determining cell death or change after the test substance is added to the culture. Probable eye irritancy is extrapolated from the results, by using information gathered in validation studies that compare the in vitro test to known in vivo data for different classes of chemicals. A growing number of companies are using these techniques to screen products for irritancy.

The **agarose diffusion** method, used by Noxell Corp. to reduce its animal use drastically, is one example of cell culture techniques. This test uses cell cultures from mice that are now available in "immortal cell lines, so no more animals are required," according to the Physicians Committee for Responsible Medicine. The committee also noted that for each product, this test takes only 24 hours and costs only about $50 to $100, compared with up to three weeks and $500 to $700 for the Draize. The agarose diffusion method has been used for testing medical devices for about 25 years but until now has not been used for consumer product testing.

Procter & Gamble is now investigating another kind of cell culture test, originally developed by Molecular Devices of California, called the **Light-Addressable Potentiometric Sensor** (LAPS), which monitors the metabolic rate of a cell culture. LAPS can also measure the cell's recovery, or lack thereof, in reaction to the test substance. This new test "has tremendous potential for automation and simplification and assesses a very general cell process, energy metabolism," said *The Alternatives Report* in fall 1989.

Still another cell culture test, known as the **neutral red uptake assay,** developed by Dr. Ellen Borenfreund and her colleagues at Rockefeller University, predicts toxicity by measuring cell damage by staining the exposed culture with a dye that is absorbed only by healthy cells. This technique is being promoted by Clonetics Corp. of San Diego.

- *Microorganisms:* One technique of this sort uses the protazoan Tetrahymena thermophila, with scientists carefully observing it after an application of test material. "If the protozoan stops moving, the chemical being tested is probably toxic enough to damage the eye," an October 1989 *Business Week* article explained.

- *Eyes from human and animal cadavers:* Whole eyes or parts of eyes from cadavers can be used to test for irritancy. In one version of this approach, described by the Humane Society of the United States, corneas from rabbits or cattle "are incubated with suspected irritants. Irritancy is inferred" from various changes that happen to the cornea. Obtaining and maintaining individual eyes from donor bodies is still a problem, however, some experts say.

- *Computers:* The Topkat program from Health Design Inc., described above, is billed as a replacement test for the Draize eye test; Topkat and other similar computer programs are used by companies to cut Draize-related animal use. In-house computer banks that contain information on previously tested compounds are very important factors companies use to reduce animal use as well, and have been the foundation for major reductions in animal use.

- *Tier testing:* The common approach to tier testing for eye irritancy is to measure the substance's pH, then to conduct cell toxicity studies with cultures, and finally to do limited in vivo tests to test the results predicted from the first two steps. At each stage, the test material can strike out if it exhibits certain criteria. (See figure below.)

Tier Testing for Eye Irritation

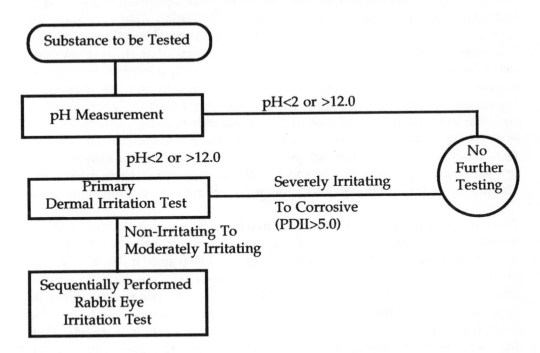

Source: Shayne C. Gad, "Acute Ocular Irritancy Evaluation: In Vivo and In Vitro Alternatives and Making Them the 'Standard' for Testing," *Benchmarks: Alternative Methods in Toxicology* (Princeton Scientific Publishing Co., 1989), p. 158.

The Draize skin test—About six rabbits are used for this skin-patch test. The hair on their backs is shaved and the skin on half of the animals is slightly abraded. Consumer Product Safety Commission regulations note that once the test substance is put on a piece of gauze and taped in place, the "entire trunk of the animal is then wrapped with an impervious material, such as rubberized cloth, for the 24-hour period of exposure." After one day, the wrapping is removed and lab workers gauge the immediate effect on the skin; after three days they make another observation and begin another round of testing on the same animals, on areas that were not previously exposed. Once the test is over, the rabbits are either killed or "recycled" for another test, as with the Draize eye test.

—*Criticisms:* Not as much intense attention has been given to the Draize skin irritancy test as to some other animal tests. General criticisms of it are that animals experience pain and suffer during the tests and that the rabbits commonly used do not produce results that can be satisfactorily applied to human experience.

—*Alternatives:* The Office of Technology Assessment notes that "even though little work has been done to develop alternatives to skin irritation tests, the many approaches...for eye irritation may eventually be applied to skin testing as well." This sentiment is echoed by other observers. The Fund for the Replacement of Animals in Medical Experiments suggests that skin patch tests for substances expected to be minor irritants be done on people, and that harsher chemicals be tested on human cadaver skin, "operation specimens," or skin cultures. The Fund for the Replacement of Animals in Medical Experiments also recommends better standardization of existing in vivo skin irritation tests. Recent advances in skin culture techniques will open the way for more specific alternatives for skin irritancy testing. By using tissue cultures to identify severe irritants, companies are starting to use human skin patch testing more extensively, cutting animal use.

Long term animal tests: The extent to which chronic toxicity tests are used to evaluate consumer products probably varies depending on the product and the company involved. In concentrating on the LD50 and Draize tests, critics of animal use in consumer product testing have not specifically targeted chronic tests that also may be conducted. IRRC spoke with several scientists to get some idea of the extent to which such testing is done. Rowan said his sense was that more animals are used in companies' acute toxicity tests than in chronic tests, because acute toxicity tests take less time and so go through more animals; although chronic tests may use 300 or more animals each, they are conducted over a much longer period, he noted. Dr. Keith Booman, technical director of the Soap and Detergent Association, told IRRC that "people wouldn't put material in a household product if it caused cancer or birth defects," but he would not speculate on what extent long term toxicity tests might be performed for household products. He stated that the decision to conduct long term tests depended on the product, and that the exposure and potency of the product must be taken into account in planning testing protocols.

Alan Goldberg of the John Hopkins CAAT told IRRC it was "hard to say" how often long term tests are conducted for consumer products, but he speculated that practices are "widely divergent." Dr. Oliver Flint of Bristol-Myers Squibb said that "most reputable companies would do large scale testing," but that the product type was the deciding factor as to what kinds of tests might be conducted. For example, he said that for a shampoo, skin, eye and allergenicity tests would probably be sufficient, but that for an insecticide, inhalation toxicity, chronic toxicity and carcinogenicity tests would be done. "Since people are unlikely to drink shampoo on a regular basis, then you don't need to do chronic toxicity tests," he concluded. Long term tests are used extensively for pharmaceutical and industrial chemical testing, however. The discussion below provides a brief look at these tests, criticisms of them and some general alternatives.

OTA described the different kinds of long term toxicity tests in its 1986 study. These tests are conducted to determine the effects a substance has on humans over time; chronic tests last a year or more, subchronic tests for around three to six months and short term tests up to a month. During these tests, which often use rats and dogs, the animals are observed and tested to see how the substance changes their behavior and physical nature; after the study period ends, scientists kill the animals and dissect them to see what has happened to them internally. Probably the best known are carcinogenicity tests. Experimenters use large numbers of rats, mice and dogs to provide statistical evidence whether a substance causes cancer, often by looking at the rate of tumor development. Experiments with animals also are used to determine possible developmental and reproductive disorders caused by substances. Many different types of tests are done in these trials, in which rats and rabbits most commonly are used, but "mice, hamsters and other mammals are used as well," OTA says. Finally, animals are used to see what may cause **neurotoxicity**. "Neurotoxicity tests would typically follow acute or chronic toxicity ones in which neurotoxic effects had been observed or were suspected," OTA reported.

—*Criticisms:* More information is gained from long term tests, experts point out, but the tests are very expensive and take a long time to complete. In cancer studies, since some species seem to be more prone to cancers than others, and because little is known about what causes cancers, animal tests are said to be imperfect models for identifying human carcinogens. Further, "The insensitivity of animal testing has long been a problem" when testing for cancer agents, Rowan said in *Of Mice, Models and Men*. Similar problems exist for reproductive toxicity tests. Goldberg and Rowan have pointed out that "The routine animal test systems have not proved particularly reliable predictors of hazards to the human fetus. For example, the human fetus is 50 times more sensitive to thalidomide than the rabbit but is totally insensitive to cortico-steroid-induced cleft palate, to which rabbits and rodents are very susceptible."

—*Alternatives:* Different methods proposed to cut animal use in long term animal studies have ranged from combining tests for the reduction of animal numbers to using cell cultures, bacteria or insects to study cell mutations and developmental disorders. The Ames mutagenicity test was one of the first in vitro tests to look at cell mutation, and its use in carcinogenicity testing today is fairly widespread. Benford noted that "significant advances have been made in [studying developmental disorders], particularly using primary cell cultures of rat embryo limb and mid brain." The National Cancer Institute has switched from a system using mice to focus on "human tumor cell lines grown directly in test tubes," the spring 1989 CAAT newsletter reported, and cell cultures are also being used for AIDS research. Computer simulations also are used.

Goldberg notes that many animal tests for chronic toxicity are empirically based, and that a better understanding of the mechanism of toxicity will help researchers predict human effects better; in vitro techniques are helping to solve these riddles. Still, a fall 1989 report from a Tufts Center for Animals and Public Policy workshop for industry representatives noted, "although it would appear that alternatives have vast potential and have received significant publicity, one participant expressed despair for the future application of alternatives in chronic toxicity testing." Because the long term effects of substances on living systems are complex, the development of alternatives for chronic tests poses a greater challenge than for more short term toxicity tests. Research continues.

Validation: Prospects for Replacement

Opponents in the animal testing debate hotly disagree on when it will be practical to end animal use for consumer product testing. Activists insist that scientific knowledge is available now to accomplish their goal. Companies dismiss this claim as untrue. Although a growing number of scientists do agree that alternatives have a potent promise to replace animal use, nearly all say total replacement is not possible right now—although it may be soon. When animal use can end depends on how fast research progresses and how soon broadly accepted validation procedures are established, scientists say. Even if a validation standard emerges from the current swirling debate, it still may be years before nonanimal tests replace all current animal methods, some scientists believe. More optimistic observers feel that once validation programs are ironed out, total replacement could be just around the corner.

The obstacles are not all scientific: OTA noted, "The implementation of alternatives is hindered by various forms of institutional inertia, such as regulatory schemes, product liability law and general resistance to change." Corporations that have come under fire for their animal use sometimes remain skeptical of alternatives too. Gillette, for example, wrote in November 1986 to activist Henry Spira that "there is no guarantee [alternative tests] will ever completely replace animal whole organ and systems testing." Regulatory agencies continue to urge caution about raising false hopes for total replacement as well. In a standard letter to the public on the issue, the Food and Drug Administration says, "It would be unwise for us to urge manufacturers not to do any further animal testing or to reject data obtained from such tests." Rowan told IRRC that there is no benefit to regulators in accepting new methods, since "if something goes wrong, they'll get it in the neck," and if the new method provides the same service as the old one, there is little change from the regulator's perspective. Adopting alternatives is therefore an "all risk, no gain proposition" for federal agencies, Rowan concluded.

Progress so far: Despite these barriers to replacing animal use completely, companies are phasing in many of the alternatives described above and cutting animal use as a result. Individual companies define their own validation standards as they choose nonanimal tests. In a November 1986 letter to Henry Spira, Avon explained its approach at that time: "By validation, we mean finding sufficient correlation with in vivo results to justify selected in-house use of the in vitro assays....We can start using the in vitro tests for product types which demonstrated good correlation while continuing to refine the technique or identify new methods for testing those product types which were not predictive....Clearly, we will not end up with one cell culture assay that will be appropriate for all products but rather a menu of assays which can be selected based on the properties of the test material." Avon now uses a combination of in vitro tests and its computer database to make product safety evaluations and does not use animals, although some of its suppliers of chemicals probably still do test on animals.

The gradual replacement of animals in testing that Avon was following in 1986 and that other companies are attempting right now clearly does depend on some kind of validation—whoever defines the standards. In a 1989 paper, Goldberg says validation is a "process which takes a test from a research curiosity to a functional, creditable methodology." He describes two basic steps: microvalidation, in which the developers of the alternative test standardize the test and conduct studies to make sure there are "no inherent biases in the test results," and macrovalidation, where different laboratories evaluate the test and a database is developed to assess the test further and store information on its use with different chemical compounds. Goldberg stresses that it "takes time to overcome the problems inherent in introducing new technologies and having them accepted." He is adamant that total replacement is not yet an option: "To eliminate animal testing at this time would constitute an abrogation of the toxicologist's responsibility to ensure safety and will pose a risk to human health that government, industry and the public will find unacceptable." Still, he says, "I have no doubt that once validation studies [are] completed, a battery of tests will be developed that will eliminate the current needs for most in vivo eye irritancy protocols." Total replacement of animals in other acute toxicity studies will probably take longer because of the many different mechanisms involved (see above for more detail), although alternative testing regimens that cut animal use drastically are likely to increase.

As scientists define the validation process further, they have to decide which evaluation standards they will employ to measure the usefulness of alternative tests. Ideally, a toxicity test would be judged by how well it predicts danger to human beings; however, there is little testing of actual human sensitivity to poisonous substances. Data gleaned from accidental poisonings is anecdotal and spotty, and not nearly as comprehensive or systematic as the

50 year store of information that uses animals as evaluation tools. A strong tendency therefore prevails to compare alternatives with the results of such benchmarks as the LD50 and Draize tests, which many scientists consider crude and occasionally meaningless for humans. Once some kind of yardstick is chosen, a test can become accepted for its consistency, sensitivity to dosage and concentration, success with a wide number of substances, and ease of application. Some scientists argue that superior safety assessment methods will be found if alternative tests go beyond correlation with the old animal-based standards and reflect the specific mechanisms of toxic reactions. Instead of using tests that just predict toxicity from correlation, like animal tests and some alternative methods, mechanistic studies could explain why the compound will react a certain way, and therefore produce a more certain forecast of human reaction. Such an approach means defining a completely new safety evaluation process, though, and implies a lengthy period of study and implementation.

The problem remains that no clearly defined, standard process exists from which existing alternative tests can receive a universal stamp of approval. Scientists who spoke with IRRC stressed repeatedly that consensus within the scientific community is built slowly. Also, "There's no procedure by which you can take a new [alternative] method and challenge the government to accept or deny a test and then say why," John Frazier of CAAT said in a 1988 CAAT newsletter. He added that even though much progress had been made in developing alternative tests, similar progress on a "strong validation program" had not occurred.

Toxicologists are concerned that validation not be rushed, since some feel hasty evaluations will only shoot down the new technology in the long run if it is not carefully evaluated and does not receive clear approval at the outset. Dr. Oliver Flint of Bristol-Myers Squibb told IRRC that if alternatives are developed too quickly, and scientists are not fully confident, the goal of reducing and replacing animal use will be "ill-served." Flint also said many claims made by the promoters of new in vitro methods may be driven by economics, and perhaps not based on solid scientific assessments. In a fall 1989 article, *The Alternatives Report* carried a similar note: "Beyond the realm of scientific investigation, the issue of economic and political promotion of certain methods has entered into the development and evaluation of alternatives. The alternatives techniques currently receiving the most attention seem to be the ones that have aggressive promotional programs." The newsletter concluded, "Are the most appropriate techniques being investigated or are efforts being spent on techniques with the largest investment?"

Some scientists also believe the pressure to placate their critics may cause companies to allocate money for alternatives research that might not improve toxicity assessments in the long run. If validation is better defined, though, researchers might be able to focus their efforts more productively, since they would have a better idea of how their current research could fit into future safety assessment programs. "We have to do something" about validation, concluded Dr. Shayne Gad, director of toxicology at G.D. Searle's research division in Illinois; "otherwise there is no value other than politics or public relations to pouring in money for more cytotoxicity tests."

As noted above, the corporations that are using emerging alternatives technology to reduce, refine and partially replace their animal use now depend on different validation standards they define themselves. With little clear guidance from federal regulators and no fully defined industry standards for validation, companies ultimately rely on their own toxicology departments to decide if they are satisfied with an alternative test. Occasional private nods from government agencies seem to indicate tacit federal approval for some specific alternative testing programs, however, as when Avon adopted Eytex and human skin patch testing in conjunction with its computer database. While some firms use new technology along with records of probable toxicity gleaned from past testing of similar products, others do not have the extensive databases needed for the latter exercise, or do not feel ready to switch over completely to alternative approaches without a specific federal mandate. For example, an official at Schering-Plough told IRRC that "it is our interpretation of the law that, in the case of the animal testing that we do, there is no adequate non-human way of testing." Further, Kevin Loftus, senior group counsel at Gillette, told IRRC that he has "never heard an attorney who is a bona fide food and drug law expert say that animal testing is not necessary," given the current state of federal regulations.

Companies are not completely isolated in their attempts to validate some existing alternatives. As described above, the Soap and Detergent Association and the Cosmetic, Toiletry and Fragrance Association are both supporting validation programs. Neither effort has given its blessing to any of the alternatives evaluated so far, however. Other work on validation has been done by the National Toxicity Program, through the Environmental Protection Agency's National Institute for Environmental Health Sciences. According to a 1986 *Science* magazine article, the program "spent about $70 million" between 1981 and 1986, "to lay a basis for validation of in vitro tests." During fiscal year 1988, the NTP allocated $17.7 million (22 percent of its total budget) to this program. *Science* further noted that "the Environmental Protection Agency plans to spend $1.5 million to validate structure activity relationships."

Setting new cogs in motion: While independent validation activities progressed at different speeds and with little central coordination, for the last few years discussions on how better to define the validation process have been an integral part of conferences on alternatives. This attention now seems to be setting some cogs in motion to establish a formal validation procedure that would have the support of federal regulators and members of the scientific community.

CAAT has established a validation committee headed by Dr. Robert Scala, a senior scientific adviser for Exxon Biomedical Sciences. CAAT's spring 1989 newsletter said the committee's "overall objective...is to catalyze the transfer of alternative technology from the research laboratory to practical applications." The committee has three specific charges: "to develop a framework to assess existing programs and coordinate future activities, to establish a scientific structure for validation [and] to maintain strong links with regulatory, academic and industrial activities." Also, Frazier of CAAT is conducting workshops to examine technical problems of validation, and is organizing a joint meeting with the European Research Group for Alternatives in Toxicity Testing. This meeting "will encompass a broad perspective on validation and will reach an international consensus on the basic requirements for this process," the fall 1989 CAAT newsletter said. Frazier hopes that these planned guidelines will be adopted eventually by the Organization of Economic Cooperation and Development, the newsletter added.

Andrew Rowan hopes to produce a similar set of recommendations in a symposium that will follow the workshops now underway with the Tufts Alternatives Project. To identify viable alternatives, Rowan told IRRC he envisions a blue ribbon panel of toxicological experts who could identify acceptable tests and give them a stamp of approval. In addition, he suggests a government database—"not just a data dump, but proactive"—that would have data storage, retrieval, analysis and structure-activity relationship programs all in one unit. This database could be used by regulators to make safety assessments of chemical compounds industry proposes to use. Much of the information needed for such a project sits in company computers now, and much of it also may be considered proprietary by these firms. Rowan's solution for this hurdle is either to guarantee a confidentiality process or to compensate companies for their cooperation in surrendering data.

Representatives from the main government agencies that deal with animal testing for safety assessments—the CPSC, EPA and FDA—have been meeting on an ad hoc basis over the last two or three years to assess the state of alternatives research and to craft appropriate regulatory responses to events, Dr. Kalish Gupta of the Consumer Product Safety Commission told IRRC. He said the suggestion for a panel of experts that would periodically assess alternative tests is a good one, but that the formal creation of such a panel under government auspices depends on the availability of funds—a proposition that may be tied more to politics than to science. An interagency group already exists informally, Gupta noted, but formalizing this group would incur some administrative costs. The real question of funding availability, however, concerns the creation of the database proposed by Rowan and others in the scientific community. Such a project has been discussed by the National Library of Medicine and the Department of Agriculture, but the issue of proprietary toxicity data held by companies is not easy to resolve. If only limited access was allowed for regulators, it might be difficult to justify the cost, while if broader access were permitted, confidentiality could be jeopardized.

It seems clear that progress on alternative tests will continue and even quicken as the discipline of in vitro toxicology gains supporters in the 1990s. The current attention focused on ironing out wrinkles and building consensus for a validation process broadly accepted by the scientific community also promises results and could speed basic research as well. What seems equally clear, however, is that for some of the more strident members of the animal protection community, this promise is not enough. Given the all or nothing approach of many animal rights proponents, continued agitation to end all animal tests is certain. This ideological commitment does not seem to impress most toxicologists working on alternatives, but it could provide the impetus for more funding—perhaps from Congress and more probably from private industry—that will make more research and validation programs possible. The flip side of increasing pressure from animal activists could also place some restrictions on animal testing, a disquieting prospect for researchers who are loath to see legislated guidelines affect their work in any way. Since the debate is bitter and many participants are deeply entrenched, though, political maneuvering may very well decide the outcome, whatever the state of alternatives science.

IV: CONSUMER PRODUCT SAFETY TESTING: COMPANY PROFILES

While boycotts and public demonstrations have played an integral part in the animal protection movement's campaigns directed at major U.S. corporations, since 1987 activists' use of shareholder resolutions has brought the debate directly to corporate board rooms and other shareholders in another way. The nine companies profiled in this chapter all have faced resolutions from People for the Ethical Treatment of Animals and have been forced to respond to questions at their annual meetings about their animal use and research into alternative, nonanimal testing methods. These nine corporations have all played an important role in the animal protection movement's publicizing of animal testing for consumer product safety. The PETA proposals have picked up a respectable amount of support in shareholder votes, although like nearly all shareholder resolutions, none has come close to receiving majority support over management's opposition.

The shareholder proposals from PETA in 1987 asked firms to halt tests not required by law, to start phasing out products that "cannot in the near future be legally marketed without painful live-animal testing" and to list which products were evaluated in tests painful to animals. In 1988, PETA dropped the request to phase out products and was rewarded by somewhat higher levels of support. Then, in 1989, the group scaled back its resolutions still further, by eliminating the request for a list of products painfully tested. For 1990, PETA is returning to all of its 1989 target companies except Avon, which had eliminated animal testing, and American Home Products, where the proposal had not gotten the necessary 6 percent support required for a resolution to come to a vote after its second year.

The table below shows the support levels for PETA's shareholder resolutions since their introduction in 1987.

	1987	1988	1989
American Home Products	—	5.5%	5.2%
Avon Products	—	6.7%	10.1%
Bristol-Myers	—	4.6%	6.5%
Colgate-Palmolive	—	9.0%	7.3%
Gillette	—	16.1%	9.9%
Greyhound	6.8%	10.4%	11.4%
Johnson & Johnson	—	—	6.7%
Procter & Gamble	2.0%	2.0%	—
Schering-Plough	—	5.8%	7.4%

PETA's resolutions have highlighted several points in the debate over corporate animal use for consumer products. Although in-house company facilities must report to the U.S. Department of Agriculture how many animals they used, and if the tests were painful, corporations do not have to say what kinds of products were used in the reported tests. Companies also are not required to report on animals that are used to test their products at outside laboratories that conduct tests under contract for them. Further, reports to the USDA need not include statistics on the number of rats and mice used—species that account for an estimated 80 to 90 percent of all laboratory animal use in the United States. PETA contends that companies must reveal all this information to the public to prove that they are making progress toward eliminating animal use.

An important concern of PETA and many others in the animal protection movement has been the pain experienced by laboratory animals, and the following profiles discuss this issue. Annual reports companies file with the USDA must say how many animals experience pain in the reported tests, but no criteria have been established as to what constitutes pain. Thus, animal protection activists say, the figures on painful tests reported are entirely subjective and could be inaccurate. If companies do report painful tests, they must provide a brief explanation as to why no anesthesia was administered.

In preparing the following profiles, IRRC asked the companies about their animal use and support of alternatives research. Some of the corporations provided information about the testing of their consumer products, as separate from pharmaceutical or health care items, that is not otherwise publicly available. Some of this testing occurred at outside laboratories. Where companies were unwilling to provide statistics on their animal use, IRRC used data from the USDA that, despite their imperfections, provide at least a limited view of the companies' animal use. When large proportions of the companies' business are not in consumer products, however, animal use data

Activists Press for 'Cruelty-Free' Companies

When animal rights activists press companies to give up animal testing, they point with enthusiasm to the companies that are now promoting "cruelty free" products. One of these is The Body Shop, a British company that does not test its products on animals, which opened stores in the United States in 1988. The company has about $125 million in annual sales and some 350 stores around the world. Also, a growing number of small cosmetics firms say their products are not tested on animals. PETA provides a list of more than 150 "cruelty-free" companies that sell such items as soap, powder, fresheners, bath oil, suntan lotion, shampoo, hair spray, toothpaste, shaving cream, deodorant, eye shadow, makeup base, nail polish, perfume and aftershave. To avoid animal testing, the "cruelty-free" companies usually test new cosmetics on human volunteers who report their reactions to scientists. Human tests can determine the efficacy and irritancy of a cosmetic and even explore likely misuses of a product, such as skin contact with detergents. These companies have been quite constrained in their product lines. No companies are marketing new disinfectants, bleaches or numerous other common household items that generally require animal tests. PETA lists another 47 companies that do not use any animal products, in addition to not testing products on animals.

"Cruelty-free" companies often formulate products from the several hundred ingredients proven to be safe, particularly "natural" ingredients such as chamomile, coconut oil, aloe vera and lecithin. At some time in the past, though, suppliers or other cosmetics firms probably conducted animal tests on these ingredients, too. For this reason, some industry representatives argue that no business really sells products without animal testing but only products that were validated by earlier tests they did not conduct themselves.

from the USDA are particularly unhelpful in providing an independent assessment of animal use for consumer product testing.

To varying degrees, all of the companies profiled have supported research into alternative toxicology tests that do not use animals. Several of the companies gave detailed information about this research.

Two of the firms profiled—Avon and Greyhound—no longer use animals to test their products. Avon announced an end to its animal use in June 1989, and Greyhound told IRRC in fall 1989 that it has declared a moratorium on animal use. All of the other companies continue to use animals for testing their consumer products, although they say that such use has dropped dramatically in the last decade as company toxicologists have started to employ some alternative methods to replace, reduce and refine their animal use. The firms still using animals maintain that alternative testing methods used alone do not yet meet their standards in determining consumer safety.

The supplier issue: After Avon made its announcement in June, some of its competitors and a few animal protection activists raised a new wrinkle in the animal use debate, which involves animal testing by the suppliers of formulas and raw ingredients. These critics expressed skepticism about the nature of Avon's stoppage, pointing out that it has not ruled out using ingredients that its suppliers have tested on animals. Initially, PETA appeared satisfied with both companies' statements, although Susan Rich of PETA told IRRC in June 1989, "We do get a little concerned when [animal testing] gets down to the ingredient level." Now, PETA is considering a renewed boycott of Avon because some of its suppliers still test raw ingredients on animals, although Avon has begun to encourage these firms to switch to alternative testing methods.

The issue of supplier relationships has entered the debate because some observers say they are reminiscent of outside testing contracts, which have been a central concern for critics of animal testing. Susan Rich, for instance, contends that Gillette "went underground" with its testing after it closed its in-house lab and started to contract out its tests.

Kevin Loftus, senior group counsel at Gillette, was skeptical of the extent of Avon's move away from animal use. He told IRRC that there is little difference between a company's relationship with an outside contractor and a supplier. "I have difficulty understanding what Avon is doing. Animals are still being used appropriately by suppliers, so what is the change?" he said. Suppliers have contracts with companies to provide a set number of product formulations, which the supplier may test on animals to provide safety data. He said the company that sells the final products may no longer either test on animals itself or contract out testing of formulations it has created, but nevertheless use products with ingredients that suppliers tested on animals.

Still, it appears that it is possible to come to terms with the supplier problem. The Body Shop, a British company, illustrates one approach for those who want to use products not tested on animals. To certify its products as "cruelty-free," the company requires regular written confirmation from its suppliers that ingredients it uses have not been tested on animals within the last five years. An additional option is for companies to use the several hundred ingredients "generally recognized as safe," instead of new materials.

Loftus said Rich was wrong to say that companies could avoid association with animal testing by using proven ingredients, arguing that a combination of ingredients already known to be safe could produce a toxic result. But Mike Dickens, an Avon toxicologist, indicated that Loftus's point was a red herring for the product safety-animal testing debate. He explained to IRRC that while a combination of safe materials can produce pharmaceutical and pesticide products that are hazardous to people, this is generally not the case for cosmetics. Toxicity data for raw cosmetics ingredients is a reliable predictor of the product of their combination, he said, and new testing is not required.

It is unclear just how many new ingredients suppliers would test on animals, since some new materials are so similar to old ones that safety data can be inferred without new animal tests. Dickens told IRRC that about 100 new ingredients come up for review by toxicologists each year. How many of these new ingredients each company uses every year is not clear, either, but Loftus said one estimate is that about 50 percent of personal care products sales came from products that were introduced in the last five years.

Whatever PETA decides to do concerning the suppliers issue, it is clear that most major consumer product companies are using fewer animals today for testing their products than they did before animal protection activists began to complain. Companies also have contributed substantial amounts of money for research into nonanimal alternatives. So far, only cosmetics companies have said they no longer need to use animals, however. The makers of other household goods argue that for their product lines, nonanimal methods are not yet sufficient to determine product safety, over the protests of activists. This chapter's company profiles take an in-depth look at what companies are saying about their continued animal use, and what kinds of research they are supporting to cut back on this use.

AMERICAN HOME PRODUCTS

Overview of AHP's operations and animal testing

Breakdown of product lines	Number of animals used in 1988*	Percent of all animals used in 1988	1988 sales (millions)	Percent 1988 sales
Health care	20,958**	97%	$4,139	75%
Consumer Products	677***	3%	$711	13%
Food Products	--	--	$650	12%
Totals	21,635	100%	$5,501	100%

Source: USDA and AHP

*   Includes rats and mice.
**   This figure combines 1988 U.S. Department of Agriculture data from AHP subsidiaries: Fort Dodge Laboratories, Franklin Labs, Sherwood Medical, Wyeth-Ayerst Research (in New Jersey and New York) and Wyeth Laboratories.
*** This figure, for AHP's household products subsidiary, Boyle-Midway, was supplied by AHP.

Most of American Home Products' animal testing is done for its health care product lines, at laboratories owned by the company. The company's household products, from Boyle-Midway, are tested by contract laboratories. These laboratories, which AHP declines to name, "are selected based on their capability in performing the necessary tests and their reputation for being in compliance with the laws governing the care, keeping and use of laboratory animals."

Household products made by AHP's Boyle-Midway include Woolite fabric and rug cleaners, Pam cooking spray, Black Flag and other insecticides, Easy-Off oven cleaner, charcoal lighter, air fresheners, Sani-Flush, and furniture and metal polishes. AHP first made public the number of animals used in its consumer product safety testing in a letter to the Securities and Exchange Commission in opposition to a 1988 shareholder resolution on animal testing. This information was updated and expanded on in the company's 1989 proxy statement. The company reported that oral toxicity tests were performed for furniture oils, a toilet bowl cleaner, pesticide products and "certain cleaning and deodorant products." The Draize test was used for furniture oils and pesticide products, while inhalation toxicity tests were conducted for a pesticide aerosol product and air fresheners.

American Home Products did not reveal how many animals at Boyle-Midway experienced pain during tests. Overall at the company, 110 animals in the USDA's "other" category experienced pain unrelieved by some kind of anesthesia.

Animals used at Boyle-Midway, AHP's household products division

	1984	1985	1986	1987	1988
Dogs	NP	0	8	20	0
Rabbits	NP	130	80	210	241
Guinea Pigs	NP	0	0	29	46
Rats	NP	122	140	330	390
Totals	NP	252	228	589	677
No. of products tested on animals	NP	14	9	35	35
Amt. spent on animal tests	NP	NP	$25,000	NP	$27,000

Source: AHP

Animal Testing Policy

AHP says it is reducing animal use with three methods: (1) comparing products with available test data, (2) using a single maximum dose, rather than several levels, in certain instances, and (3) using low toxicity ingredients. "Wherever possible," the company told IRRC in 1988, "single dose limit tests are used in place of the multi-dose LD50 test." The company also has avoided Draize eye tests when the skin irritation tests are positive, on the assumption that eyes also would be irritated.

AHP says it does not conduct tests when data are available for a very similar product, when toxicity can be inferred from the chemical structure, or where the Environmental Protection Agency will accept toxicity data on the active ingredients of a product in lieu of the whole product.

Nonetheless, product development at Boyle-Midway requires animal testing, AHP says. The company told the SEC in 1988 that most of Boyle-Midway's products "are complex chemical formulations, and frequently require re-testing whenever an ingredient is added, removed or the mixture is altered, since the result could be a new level of toxicity, irritancy or corrosiveness that is unknown and unable to be established by existing data." For example, "the mere addition of a fragrance results in an innocuous change in the product mixture," the company wrote, but "it can so alter the toxicological effect of the entire combination of ingredients" that testing is needed. One change, according to AHP, "can be significant enough that there is a whole new product, and that product must in one way or another be tested for its safety."

Alternatives Research Program

AHP told IRRC that "Boyle-Midway has initiated a program to comprehensively explore alternative test systems to find those that can be used to reduce the use of animals, to reduce their pain and suffering and to replace them, if possible, and still provide data that is acceptable to the various regulatory agencies with jurisdiction over Boyle-Midway's products." The company said it has also supported research on alternatives through trade associations and by donations to the Johns Hopkins Center for Alternatives to Animal Testing.

AVON

Overview of Avon's operations and animal testing

Breakdown of product lines	Number of animals used in 1988*	Percent of all animals used in 1988	1988 sales (millions)	Percent 1988 sales
Beauty Products	2,423**	100%	$3000	100%

Source: Avon

*   These figures illustrate the last full year of Avon's animal use before the
     company switched to all nonanimal testing methods.
**   Provided by Avon; includes rats and mice.

Avon was the first major U.S. company to announce the end of animal testing
for its products, in June 1989. This profile describes the company's last full
year of animal use and the route it took to ending that use.

Since Avon sells only cosmetics and other beauty products, all of its animal
testing fell under the definition of consumer products used in this report.
Avon conducted all of its product tests at in-house facilities until 1988, when
it began to contract out what it termed "the relatively limited amount of testing which continues to be necessary and prudent." This move was characterized in the company's 1989 proxy statement as "an interim measure of progress." In a letter to IRRC, the company said the 1988 figures it provided to
IRRC "include outside contractors since we no longer have an in-house animal
facility."

Confirming the company's move to outside testing, PETA provided IRRC
with documentation showing that Avon sent two formulations to the Biosearch testing laboratory in Philadelphia, in June 1988, for tests using the
Draize method. PETA said its documentation suggested that Avon might
have used Biosearch on a more regular basis, and it pointed out that the
Philadelphia facility had been the subject of an investigation for animal welfare violations. The company told IRRC that the formulations sent to Biosearch were "standards"—products that already had been tested at Avon labs.
These trial tests were sent to many labs as part of Avon's effort to pick outside

laboratories to do its product testing once it stopped its in-house program in the spring of 1988, the company said. Avon told IRRC that it ultimately decided not to contract out its testing to Biosearch and instead used other labs, which it would not name.

As can be seen on the table below, animal use at Avon dropped dramatically over the last decade. The 2,423 animals used in 1988 comprised just 17 percent of the 14,550 used in 1981. The company used animals from January to June of 1989, so 1990 will be the first full year in which Avon does no animal testing.

Data from the USDA provide information on how many animals reportedly were not given anesthesia in painful experiments; they show that at Avon no painful tests were administered without anesthesia in 1987, and that in 1986 two rabbits underwent painful experiments without relief. Further, Avon told IRRC that no animals underwent painful tests without anesthesia during 1988.

Animals used at Avon

	1981	1982	1983	1984	1985	1986	1987	1988
Rabbits	3,519	2,880	1,547	1,183	848	621	477	332
Guinea Pigs	3,393	2,586	3,057	2,176	1,948	1,816	1,323	727
Rats	7,045	3,451	2,272	2,523	1,371	1,291	1,283	814
Mice	593	500	440	1,126	881	987	980	550
Totals	14,550	9,417	7,316	7,008	5,048	4,715	4,063	2,243

Source: USDA and Avon

Product Testing Policy

On June 22, 1989, Avon Products became the first major U.S. cosmetics company to announce an end to the use of animals for product safety testing. Since 1981, the company has consistently reduced its animal use, conducted in-house research on nonanimal alternative tests, and supported the efforts of the Johns Hopkins Center for Alternatives to Animal Testing. This record proves that Avon's June announcement was the culmination of many years of planning, the company said, but People for the Ethical Treatment of Animals heralded the step as a major victory that it attributed to its own four-month-old boycott to "stop Avon killing."

Avon said it eliminated its use of the Draize eye irritation test in February 1989 by substituting the Eytex system, a biochemical test that uses no live animals, which was developed by the National Testing Corp. in California. Eytex "measures a material's ability to destroy or bind to proteins and substances similar to those found in the eye," the company told IRRC. Later in 1989, the company replaced Draize skin irritation tests with human skin patch testing. However, "the foundation of the new safety program is the extensive computerized database [that] contains historical testing data on Avon's products and ingredients that the company has developed over the years," a company press release explained. Using the database, company toxicologists can compare new formulations with previously tested Avon formulas to determine the new products' safety.

Avon's route to nonanimal testing: In 1986 the company published a brochure about animal testing that asserted: "Avon's objective is to eliminate animal tests for product safety." The brochure said Avon used the limit test for oral toxicity studies that involved fewer animals than the classical LD50 method. In a 1986 letter and report to the cosmetics industry, Avon explained that its limit test used five animals for a single dose and "only rarely" required an additional five animals for a second dose.

In Avon's work on an alternative to the Draize eye test, it initially modified the method by applying a local anesthetic, diluting shampoos and other irritating products to 25 percent of full strength, holding the rabbits rather than placing them in stocks when putting the test ingredient in the eyes, and not testing substances known to be strong irritants. An Avon toxicologist told IRRC in 1988 that a company biostatistician had determined that the number of rabbits needed for a given Draize could be reduced from six to three and continue to provide the necessary level of judgment. Avon's Draize test did not blind rabbits, the toxicologist said in 1988; he mentioned that he had demonstrated the test to Avon's chairman during a tour of the laboratory.

By early 1989, Avon's use of nonanimal techniques and its database enabled company scientists "to approve more than 90 percent of our new product formulas without any animal testing," Avon said. The company told IRRC that most final formulations did not need to be tested because Avon had a sound scientific rationale for approval, such as sufficient safety data on key ingredients. With many products, an Avon official explained to IRRC, the laboratory bypassed animal studies and went directly to testing cosmetics on humans for use and skin reactions.

Avon said it was able to switch to the Eytex method after company scientists validated the method by "testing 154 products and ingredients for which Avon had Draize test data." The company evaluated the test for its ability to "accurately predict eye irritation" in "each product category"; the test's performance was acceptable for "most but not all product categories." Instead of waiting to find one nonanimal method that would work for all its products, the company said its approach was "to define where an alternative test would work and use it within those limitations." Because of this strategy, it was "able to quickly adopt Eytex within defined formulation boundaries based on the positive results. Additional alternatives are presently being evaluated to cover the remaining product categories," the company told IRRC in a recent letter. Avon shared its experience using Eytex with the scientific community at the spring 1989 conference at CAAT, and it also showed its results to the FDA.

For judging skin irritancy, Avon now uses a combination of alternative methods: First, a judgment about the new formula's probable irritancy is made by using the company's "extensive computer database." Then, human volunteers are used in skin patch testing "to confirm the mildness of the product."

The company told IRRC that, "If its new nonanimal approach is ever insufficient to substantiate the safety of a new product, Avon would just not market that particular product."

The company was unable to break down its testing patterns over the years into a numerical comparison between the number of products tested on animals and the number tested using alternative methods, or by the dollar amount spent for the two types of testing. "We only tracked the percentage of formulations submitted to Toxicology that were approved without animal testing for the last few years. This percentage remained over 90 percent since 1987," Avon told IRRC.

Support for Alternatives Research

Avon is unusual among consumer products companies in that in 1986 it opened an in-house cell culture laboratory to validate selected in vitro assays for predicting irritation. The laboratory took a wide range of cosmetic products, already assessed through animal testing, and determined whether the results of in vitro tests would correlate with the animal data. In a letter to activist Henry Spira describing the project, Avon wrote that the laboratory's purpose was to end up with a menu of assays that could be selected according to the properties of the test material, and not to find a single test that worked on all products. Avon's search for alternatives involved the use of animals; the company attributed the use of 424 (43 percent) of its mice in 1986 to experiments that were conducted in vitro. In 1988, Avon asserted in a letter to IRRC that "Ours is an aggressive program of pushing science forward as quickly as possible while still maintaining the integrity of our safety program."

The company's monetary support over the years to alternatives research outside its own laboratories comes to at least $805,000 since 1981. Avon is one of the major contributors to the Center for Alternatives to Animal Testing at Johns Hopkins University. In 1981, following pressure on cosmetics companies from animal rights activists, Avon committed $250,000 to create CAAT and announced that it had "set aside an additional $500,000 for specific research projects that will evolve from the work" of CAAT. In mid-1988, Avon committed itself to a four-year, $300,000 grant to CAAT for work on finding alternatives to the guinea pig skin allergy test. The company has also funded a similar research project in England, the Fund for the Replacement of Animals in Medical Experiments; to date, it has given approximately $255,000 to FRAME. In addition, "We have other private grants/agreements with individual investigators and new technology companies for which we do not release financial data," the company reported to IRRC.

Avon has indicated to IRRC that it would continue to conduct research into nonanimal alternative tests, since many more methods are needed "to cover all aspects of toxicology."

Controversy over Suppliers

While Avon's pledge not to use animals covered its product testing at both in-house and outside laboratories, the company left open the possibility that some of its suppliers of chemicals might still conduct animal tests on raw ingredients provided to Avon. Dr. Mike Dickens of Avon's toxicology department told IRRC that the company would accept safety data from animal tests if it is provided by suppliers, but that the company would not require or ask for such data. In July 1989, the company began to encourage its raw materials suppliers to move to nonanimal tests. PETA, however, has construed the suppliers' testing as possible grounds for a renewed boycott, although some other animal protection activists have criticized PETA for taking this stance. (See the introduction to this chapter for more on suppliers' testing and Chapter I for more on the activists' disagreement.)

BRISTOL-MYERS SQUIBB

Bristol-Myers merged with Squibb in 1989, forming Bristol-Myers Squibb. Squibb produces no consumer products under the definition used in this report, however, and in an effort to provide a consistant picture of animal use at Bristol-Myers, the figures and information used in this profile do not include Squibb. For those interested in an aggregate picture of the new company, historical animal use data from Squibb are included at the end of this profile.

Overview of Bristol-Myers's operations and animal testing

Breakdown of product lines	Number of animals used in 1988*	Percent of all animals used in 1988	1988 sales (millions)	Percent 1988 sales
Health care	5,379*	97%	$4,142	69%
Consumer Products	166*	3%	$1,831	31%
Totals	5,545	100%	$5,973	100%

Source: USDA, Bristol-Myers Squibb

*Estimate; Bristol-Myers Squibb provided only percentages for its animal use; also does not include animals used at outside laboratories, rats or mice.

The vast majority of Bristol-Myers's reported use stems from testing and research for its health care products. Although the company conducts much of its testing at in-house facilities, an undisclosed portion of product testing is contracted out. Bristol-Myers Squibb did not provide IRRC with the number of animals used in testing at outside laboratories, but it said its "use of outside laboratories remains essentially constant from one year to the next." The company also did not provide animal use statistics for the testing of its consumer products, as separate from its other business, although it told IRRC that in 1988 just 3 percent of its animal use was for consumer products, and in 1989 this number dropped to 1 percent.

The company strongly expressed its opinion that information on its animal use gives "no real indication of a company's commitment to socially responsible laboratory animal use," because of 1) increases and decreases in use "attributable to acquisitions and divestments," 2) the "level of research conducted from one year to the next," 3) "the varying test requirements from one type of research or product to another," 4) "testing requirements for new and existing products entering foreign markets," and 5) "changes in testing regulations for new and/or existing or improved products." A more "holistic" view of the company's commitment to responsible animal use comes from an assessment of its efforts to "reduce reliance on animal testing methodologies and assuring the humane treatment of those animals necessary, rather than the absolute number of animals used," the company wrote in a letter to IRRC in late 1989.

Bristol-Myers Products was the only consumer products segment of the company to submit a separate report to the USDA for animal use, although two other segments of the company—Clairol and Drackett—also make consumer products. Clairol makes hair coloring and hair care items, and Drackett produces items such as Windex glass cleaners, Drano drain openers, Renuzit air fresheners and Endust cleaning aids. The company reported to IRRC that "nearly all of Clairol's testing is done in-house at...Bristol-Myers Squibb facilities"; it did not provide information on where testing for Drackett products is conducted.

A breakdown of the company's animal use according to USDA reports produces proportions different from those reported by different product categories to IRRC by the company, because it does not account for the outside testing of consumer products, nor for the testing of consumer products done at company facilities other than Bristol-Myers Products. The 562 animals used at Bristol-Myers Products comprise 10 percent of the company's total reported animal use for 1988 of 5,545 animals. Since Bristol-Myers Products does make some over-the-counter medicines, such as Bufferin, Nuprin, Excedrin, Comtrex and No Doz, in addition to items such as Vitalis men's hair care preparations, it does not fall neatly into a nonmedical-related category for consumer products.

Nevertheless, the USDA data from Bristol-Myers Products give the only available information on at least some animal testing for nonmedical consumer products at the company, even though they include the over-the-counter medicines noted above. The table below therefore shows only partial figures for animal use for consumer product testing at Bristol-Myers. The company declined to provide IRRC with numbers that more precisely described its animal use for consumer products.

At Bristol-Myers Products during 1988, 27 rabbits underwent painful tests that were unrelieved by painkillers. The company stated on its report to the USDA: "These animals were used in tests to determine [the] skin irritation potential of product ingredients. Use of systemic or local acting drugs would compromise test results by altering blood distribution and tissue response.

Further, by subjective measurements such as body posture, weight gain, grooming, and other activities, the pain or distress was of low intensity and of short duration."

Animals used at Bristol-Myers Products

	1984	1985	1986	1987	1988
Dogs*			81	28	35
Rabbits			506	371	322
Guinea Pigs	Not		306	198	205
Frogs	Provided				181
Totals			893	597	743

Source: USDA

*   Bristol-Myers Squibb noted that its use of dogs at Bristol-Myers Products was not for consumer products, and that its outside laboratories also did not use dogs for testing its consumer products.

Animal Testing Policy

Bristol-Myers Squibb says it does not conduct animal tests when the safety of an ingredient is already known from its own or other databases. The company told IRRC that it is company policy "to insist that its scientists *always* consider replacing the use of animals by other methods; reducing the number of animals used when such tests are necessary; and refining procedures in order to minimize distress, pain and suffering. Our scientists are held individually accountable for compliance with this policy."

Animal welfare: Bristol-Myers established a permanent subcommittee on animal welfare in 1982, which is to supervise animal use at in-house and contract facilities. Animal care guidelines include "internal audit mechanisms to ensure their strict enforcement," and are "more rigorous" than those at the National Institutes of Health, according to the company.

A company committee reviews animal testing protocols before implementation. Company employees who handle animals have all viewed a video that tells about the company's policy on animal care and use, receive an in-house newsletter, *Lab Animal Topics*, and participate "in programs sponsored by such professional groups as the Scientists Center for Animal Welfare and the American Association for Laboratory Animal Science," Bristol-Myers Squibb told IRRC.

Also, Bristol-Myers Squibb said, it gave a $5,000 grant to a Tufts professor to help produce what the company described as the first-ever text on veterinary ethics, published in 1989.

Animal testing: "Safety testing of new consumer products does not routinely use laboratory animals. Unless specifically required by law (domestic or foreign), animal testing is employed only after all relevant nonanimal methodologies have been exhausted and an unacceptable level of doubt remains concerning the possibility of a significant adverse human reaction," the company told IRRC.

The company said in its 1988 proxy statement that it conducts a "limit test" for acute oral toxicity that involves five or 10 animals, whereas classical LD50 tests use up to 100. Bristol-Myers also told IRRC in 1988 that in the early 1980s it established in vitro screening capability for mutagenicity, which it uses to eliminate potential products before they are tested on animals. The company said: "Raw materials or products that are expected to be severe eye irritants or corrosive to the skin are so labeled without recourse to animal tests."

A draft of this profile sent to Bristol-Myers Squibb included a chart in which the company could show the numbers of products tested on animals and the amount spent on research on alternatives from 1984 through 1988. The company told IRRC that it "does not make public the information requested."

Alternatives Research Program

Bristol-Myers conducts in-house research on alternatives to animal testing and support outside efforts for in vitro research, as well. In 1983, the company established an investigative toxicology unit that has developed various biochemical and cellular screens to test pharmaceutical drugs, and "was one of the first U.S. pharmaceutical companies" to do this, it said. In vitro screens have reduced the numbers of animals used in drug development but still are not capable of replacing animals in consumer product safety tests, the company reported.

Alternative testing methodologies are promoted by several company committees and task forces and through "membership in and monitoring of relevant organizations," the company said. In an effort to facilitate intra-company alternatives research, Bristol-Myers held a meeting attended by more than 100 of its toxicologists from around the country in late January 1989. This event, "an annual interdivisional conference on in vitro methods...serves as a catalyst for the integration of new and improved methods into support and research programs." The company told IRRC, "It is worth noting that the annual conference was initiated by the scientists themselves."

"We are not a newcomer to the search" for alternatives, the company reported. "Historically," the company wrote, "our use of in vitro tests has spanned product safety, discovery and development methodologies." Bristol-Myers Squibb listed 26 in vitro test methods it uses. The manager of the company's biochemical and cellular toxicology department, which is "dedicated to the development of in vitro methodologies for screening new drugs for toxicity," helped to establish the Fund for the Replacement of Animals in Medical Research in the United Kingdom. "Company scientists regularly publish and make presentations on their experience with in vitro methods," Bristol-Myers Squibb said.

Currently, Bristol-Myers Squibb told IRRC that it is collaborating with a small biotechnology company, Marrow-Tech, which has developed a "proprietary tissue culture system [that] has a number of potentially significant therapeutic and laboratory testing applications. To date, the company has successfully grown bone marrow, skin and liver tissue. This cooperative effort will further reduce our reliance on animal testing methods."

Finally, the company reports that its subcommittee on alternative methods "evaluates in vitro methods for their applicability to our [consumer] product lines. Work in progress includes validation testing for a number of commercially available in vitro methods using a cross section of chemicals commonly used at Bristol-Myers Squibb."

In addition to its in-house work, the company is one of the leading funders of other in vitro research for product safety tests. It says it has donated more than $1 million for research on alternatives since 1981, including $800,000 to the Center for Alternatives to Animal Testing and $150,000 to the Fund for the Replacement of Animals in Medical Experiments in the United Kingdom. In 1987, CAAT announced that it would receive a gift of $200,000 over a three year period from Bristol-Myers.

Bristol-Myers Squibb is a member of several trade organizations that support alternatives research. *The Alternatives Report,* a new newsletter from the Tufts Center for Animals and Public Policy, is partially funded by the company. Further, the company says it is an "active participant" in the Tufts Center's project to develop a plan to implement in vitro testing techniques.

Animal Use and Alternatives Research at Squibb

In response to inquiries about historical information on Squibb's support of alternatives research and use of nonanimal tests that might be applicable to the merged company's consumer products testing, Bristol-Myers Squibb told IRRC that its "program for the responsible use of laboratory animals...applies to the company as a whole."

Overview of Squibb's operations and animal testing

			Animal Species Used				Total	Sales
Year	Dogs	Rabbits	Primates	Guinea Pigs	Hamsters	Other	Animals	(millions)
1986	842	1,646	503	3,232	331	216	6,770	$1,785
1987	827	1,280	242	2,091	150	47	4,637	$2,157
1988	600	1,708	338	2,829	200	437	6,112	$2,586

Source: USDA, 1988 Squibb annual report

COLGATE-PALMOLIVE

Overview of Colgate-Palmolive's operations and animal testing

Breakdown of product lines	Number of animals used in 1988*	Percent of all animals used in 1988	1988 sales (millions)	Percent 1988 sales
Health care **	21,915	99%	NP	NP
Consumer Products***	224	1%	NP	NP
Totals	22,159	100%	$4,734	100%

Source: USDA, Colgate-Palmolive

*   Includes rats and mice.
**   In October 1988, the company sold its health care division, Kendall; Kendall laboratories nevertheless used animals during 1988 and reported this use to the USDA. Animal use reports to the USDA are submitted for October-September, so these animals were used while Colgate still owned Kendall.
*** Includes oral care products, which account for 10 percent of the company's consumer products sales.
NP Not Provided.

In response to a request for updating of this profile, Colgate-Palmolive told IRRC that "use of animals has been reduced more than 90 percent during the past five years." The majority of Colgate-Palmolive's animal use during 1988 was for health care products produced by the company's Kendall division, which Colgate sold in October 1988. The company's product lines now are mostly nonmedical consumer products, although it still produces some over-the-counter and prescription drugs, as well as oral care products. Colgate's consumer products testing occurs in part at an in-house facility, but an undisclosed number of animals are used at laboratories that contract with the company.

The company has declined to provide IRRC with data from testing conducted at outside laboratories, although since 1987 it has given IRRC its reports to the USDA before they are publicly available from the government. A Colgate official told IRRC in response to requests for information for the 1989 shareholder resolution on animal testing that "we know exactly how many animals we use, and require a report after each assay" on test results and the animals' reactions. Although the company keeps detailed individual records of tests for tested formulations, it does not add up all the animals that are used, the official explained. He said the number of animals tested at outside labs is "very small."

The company's household and personal care products accounted for 90 percent of its sales in 1988. Colgate has a worldwide toothpaste market share of 41 percent and also makes soaps, shampoos, deodorants, dishwashing products, cleaners and detergents. Products are marketed under such brand names as Colgate, Ultra Brite, Palmolive, Ajax, Fab, Spray 'n Wipe, Irish Spring, Halo and Rapid Shave. Colgate also has a few other lines of business; its 1988 annual report said its Hill's Pet Products has the largest veterinary staff of any pet food company in the world.

A Colgate official told IRRC in 1988 that no painful procedures for consumer products had been conducted in recent years by the company or its outside contractors without the application of anesthesia. He said that the 12 cats used in 1987 and 1988 were for testing kitty litter and that homes with company employees were found for the animals after the tests. The dogs are part of a "permanent colony" and are used for tests on fluoride, plaque and tartar control formulations, he said, since their teeth and saliva are similar to humans'. Toxicity tests are conducted on the rabbits, the company told IRRC, but it said the number of rabbits used has been reduced because of the company's development and refinement of in vitro toxicity tests.

Animals used for consumer product testing at Colgate-Palmolive

	1984	1985	1986	1987	1988
Dogs	40	50	53	55	57
Cats	0	0	0	12	12
Rabbits	657	232	151	253	175
Hamsters	240	278	11	0	0
Guinea Pigs	538	721	292	0	0
Totals	1,475	1,281	507	320	244

Animal Testing Policy

Colgate says it does not conduct tests when toxicological information is available from its own databank, ingredient suppliers or outside sources. When animals are used, tests are "conducted under controlled conditions with sensitivity to animal welfare considerations." Colgate-Palmolive's director of pharmacology and toxicology approves and oversees in-house and contracted animal testing, the company says. A Colgate official told IRRC that the company formally wrote up a policy on animal testing in 1988, although it existed in practice before. The policy states that "contracted animal research is subject to comprehensive field audit" at the discretion of the director of pharmacology and toxicology. The company also has modified its Draize eye irritancy test and cut back on the number of animals needed for oral toxicity studies, Colgate reports.

In preparing the draft profiles for this report, IRRC included a chart in which each company could show the numbers of products tested on animals and the amount spent on research on alternatives from 1984 through 1988. Colgate returned its profile uncorrected, with the chart left blank, but with a three-page statement on its work on alternatives.

Alternatives Research Program

Colgate-Palmolive's work on alternatives has been extensive. One of the company's key in-house efforts to find nonanimal alternatives has been to determine eye irritancy by testing substances on chicken egg membranes instead of on the eyes of rabbits with the Draize test. Company scientists have published studies on this test—the CAM assay—which a Colgate official told IRRC is "used increasingly to substitute for the Draize test." The CAM assay test has been further improved in the last year, the company recently told IRRC, although it is still "not a complete replacement for the Draize test." Colgate notes, "An expanded database using [the newly revised CAM assay] is being developed by Colgate internally and in external validation programs." Known as the CAMVA, the improved CAM test was presented to federal regulators in Washington in early April 1989. A company official told IRRC that the government regulators were "delighted" with Colgate's progress on the CAMVA. He told IRRC that it was still too early to tell if any of the regulatory agencies would make some kind of new policy statement indicating broader acceptance of product safety data from in vitro tests, however. Colgate's statement on alternatives says the "CAM assay is currently being evaluated in both the Soap and Detergent Association and the Cosmetic, Toiletry and Fragrance Association round robin studies."

In addition to the eye irritation work that led to the CAM assay, Colgate told IRRC in late 1989 that it has done work on developing in vitro assays for skin irritation based on the utilization of human skin cells in culture. It said it had worked with the Clonetics Corp. and Marrow-Tech to develop the models.

Colgate says it "strongly supports and funds a number of research programs directed toward the development of alternative test procedures." The company participates in university programs, trade association efforts and external research groups. In 1987, the company established an annual two-year post-doctoral fellowship through the Society of Toxicology for research fully directed towards alternative methods of testing that can remove the need for animals. The company says the first two fellows "are investigating relationships between various in vitro models and in-vivo dermal effects. The work will advance the progress to reduce the use of animals for skin irritation testing."

The Johns Hopkins Center for Alternatives to Animal Testing lists Colgate as a "corporate benefactor," which is a category for donors of $5,000 or more. In addition, Colgate conducts its own research and, it says, "has allocated substantial space and employed scientific professionals to find and develop new alternative testing methods." A company official told IRRC that Colgate has spent "several million dollars during the last few years for internal and external programs" aimed at reducing and replacing animal use. The statement it sent IRRC on alternatives in late 1989 said its principal 1989 outside donations, in addition to CAAT, had been to the Department of Dermatology, Columbia University College of Physicians and Scientists, which is working with Colgate on the in vitro assays for skin irritation.

GILLETTE

Overview of Gillette's operations and animal testing

Breakdown of product lines	Number of animals used in 1988*	Percent of all animals used in 1988	1988 sales (millions)	Percent 1988 sales
Consumer Products	(outside lab)	100%	$3,581	100%

Source: USDA and Gillette

* Animal use figures for Gillette are not available for 1988 because the company closed its in-house testing facility during 1987 and has declined to provide IRRC with subsequent animal use data.

Although Gillette makes some over-the-counter and prescription drugs, the majority of its products fall into the consumer products category used in this report. Since 1987, all of Gillette's animal testing has been done by outside laboratories. Before 1987, Gillette conducted most of its own product safety testing on animals at the Gillette Medical Evaluation Laboratories in Rockville, Md. In December 1986 the company decided to dispose of a number of properties and less profitable businesses in response to a takeover attempt by Revlon. The company also committed itself to reducing its work force by 8 percent, or 2,400 people, worldwide. As a result, the company sold its Rockville facility, which housed GMEL, and discontinued GMEL's in-house animal testing program. The GMEL decision occurred less than six months after controversy erupted over its treatment of animals in September 1986 (see below), and animal activists claimed responsibility for the closing.

In addition to its razors and blades, the company makes such toiletries and cosmetics as Right Guard deodorant, Foamy brand shaving cream, and Aapri and Jafra cosmetics. Gillette also sells Liquid Paper correction fluid, Paper Mate and other writing instruments.

Partial information on the company's history of animal usage is available from reports to the USDA. Gillette's reported use from 1977 to 1986 ranged between 1,919 and 4,208 animals; usage for the last five years is highlighted in the table below. The 1987 total covers the period from Oct. 1, 1986, through

March 13, 1987, when animals were no longer tested at the facility.

During 1987, the company reported that four rabbits were used in painful tests without any painkillers. A company representative wrote IRRC that "more than 150 rabbits were used in standard occular and dermal toxicity tests. Ordinarily such tests do not cause pain. However, the reactions to the test materials were stronger than usual with these four rabbits, so it was assumed that the rabbits must have suffered pain. Accordingly, the company conservatively reported the instances in the required USDA annual report."

Animals used at Gillette

	1984	1985	1986	1987	1988
Primates	4	0	0	0	NP
Rabbits	1,600	1,467	1,349	155	NP
Hamsters	2,170	1,304	869	111	NP
Guinea Pigs	434	295	285	155	NP
Totals	4,208	3,066	2,503	421	NP

Animal Testing Policy

The company reports that it has taken several steps to reduce its use of animals. The number of animals was lowered through use of a computerized database of past testing by reducing the number of animals needed for each acute toxicity and eye irritancy test, and by using the same rabbits for both the eye and skin irritancy studies. Also, medical review officers at Gillette "routinely screen all proposed ingredients through a computerized safety information database and reject those which are unsuitable...." The company told IRRC that it had modified the Draize eye irritancy test by administering a local anesthetic to the rabbit's eye before testing. "In some cases, Gillette has been able to limit—or even forgo—the use of animals through the literature and data research," it stated. The 1988 proxy statement said that "more than 60 percent of the Gillette products that were given medical safety clearance during the past five years involved no animal testing." A company official told IRRC that the statement remained true in 1989.

Gillette maintained a laboratory animal care committee at GMEL to monitor animal use and assure their humane treatment. At the time, Gillette said, "This committee has initiated control and authorization procedures which are almost identical to those used by committees which review and approve human clinical research."

In preparing the draft profiles for this report, IRRC included a chart in which each company could show the numbers of products tested on animals and the amount spent on research on alternatives from 1984 through 1988. Gillette returned its profile with the statement above the chart that "The company was unwilling to provide this data."

Alternatives Research Program

Gillette says it is currently conducting tests with the Eytex System, a screening alternative to the Draize eye irritancy test. This effort marks a return to in-house in vitro research by the company that was abandoned in late 1986 because the research had been fruitless, according to company officials.

Gillette has said it has been involved in research on alternatives to animal experimentation. In 1983, the company reported that it was one of four industry laboratories initiating a program to modify eye safety evaluation. It also reported that it participates in the Cosmetic Ingredient Review Program, which the company says is sponsored by industry and sanctioned by the FDA, as well as "CTFA (Cosmetics, Toiletries and Fragrance Association) efforts to modify existing animal testing techniques to minimize discomfort to the animals." The company said it supports the Center for Alternatives to Animal Testing at Johns Hopkins, but it has been unwilling to specify the level of its support.

Controversy over Animal Use at Gillette

As discussed in Chapter I, one of the major events that helped to provoke controversy over animal tests for consumer products occurred at Gillette. During fall 1986 Ark II, an animal rights group, called an international boycott of Gillette to force Gillette to end its use of animals in consumer product tests. An unusual undercover operation prompted the Gillette campaign. A lab technician named Leslie Fain gathered information on animal experiments during a year and a half of work at Gillette, kept a diary and videotaped actual tests and handling of animals. Based on Fain's inside view, Ark II questioned Gillette's claims about improvements in testing methods and charged that the company mistreated its animals.

Gillette denies Fain's charges and disputes her credibility. The company told IRRC at the time of the Ark II allegations that "statements and video clips in recent news releases are grossly misleading." It said, for example, that a videotape shown at Ark II press conferences showed a number of "stock footage" scenes with no acknowledgement that this was not taken at Gillette. Gillette told IRRC that "None of the most disturbing photographs of damage to rabbit eyes even involved Gillette testing. Rather, they were slides and photographs published in a U.S. Consumer Product Safety Commission manual."

In February 1987, the USDA's Area Veterinarian in Charge completed a thorough investigation that was conducted because of Fain's allegations and the resulting publicity. The investigation did not show the extent of violations alleged by Ark II, but found that some handling violations existed, and a warning letter was issued to Gillette.

GREYHOUND

Overview of Greyhound's operations and animal testing

Breakdown of product lines	Number of animals used in 1988	Percent of all animals used in 1988	1988 sales (millions)	Percent 1988 sales
Consumer Products	282	100%	$962	29%
Other product lines	0	0%	$2,343	71%
Totals	282	100%	$3,305	100%

Source: USDA and Greyhound

All of Greyhound's animal testing has involved consumer products made by the company's Dial Corp. subsidiary. Greyhound wrote IRRC on Nov. 9, 1989, "that for almost one year it has conducted no tests either in-house or at outside laboratories. On May 1, 1988, the company closed its animal testing facility at Dial's Scottsdale, Ariz., Technical Center." The company said it has declared a moratorium on all animal testing and is asking its suppliers to provide it with ingredients that are already approved as safe for use in products.

The Dial Corp. incorporates what had been the Armour-Dial and Purex companies. Its main products are Dial soap, shelf-stable food products, floor care products, laundry and household cleaning products. With well-known brand names that include Dial anti-perspirant, Pure & Natural soap, Purex bleach, Parsons' Ammonia, Dutch brand cleaners, Brillo pads, La France whitener, Trend detergent and Tone soap, Dial sells nearly 20 percent of all soap in the United States. In 1988 Greyhound acquired the 20 Mule Team, a division of U.S. Borax and Chemical Corp. that makes household products and industrial specialties.

The company says that the number of animals used in testing has declined each year since 1976 and that animal use fell by 64 percent from 1987 to 1988 and has fallen by 78 percent since 1984. During 1988, 84 rabbits and 10 rats were used in tests with no pain relief. The company stated on its report to the USDA that

> Rabbits were used to determine the toxicity of new chemicals and product formulations to dermal and ocular tissue....Anesthetics, analgesics or tran-

quilizers are not usually administered to rabbits....The rabbits may have experienced some stress as a result of the wrapping procedure used to assure skin contact with the product being tested.

The objective of the eye irritation test is to evaluate the potential for a test material to cause damage when it accidentally comes in contact with the human eye. All rabbits used in eye irritation studies without administration of anesthetic, analgesic or tranquilizer drugs were exposed to a low volume of the test material as opposed to the 0.1 ml or 0.1 g specified by the Draize test protocol. One can assume that since a lower volume of test material will produce less ocular reaction that the animal will experience less pain or distress than would be produced by a larger volume of test material producing a greater ocular reaction.

Concerning the 10 rats that underwent painful tests, the company wrote:

Rats were used to determine the acute toxicity of materials upon oral ingestion. The objective of the rat oral limit test is to [remainder of sentence deleted by USDA's FOIA staff]. Discomfort might have been produced briefly in these animals when the test material was introduced into the stomach by use of a ball-tipped intubation tube.

Animals used at Greyhound

	1984	1985	1986	1987	1988
Cats	10	0	0	0	0
Rabbits	238	202	216	167	96*
Guinea Pigs	166	147	75	70	0
Rats	630	371	342	425	150
Mice	235	86	324	109	36
Totals	1,279	806	957	771	282

Source: USDA and Greyhound

* Includes 6 rabbits tested at an outside facility in one test.

Animal Testing Policy

Before it declared its moratorium on animal testing, the company used literature searches and a company database to avoid repeating relevant tests; it used the limit test rather than classical LD50 to determine acute toxicity. Dial told IRRC that it calls on a databank of information to find ingredients that have already been tested and therefore do not require additional testing.

In preparing the draft profiles for this report, IRRC included a chart in which each company could show the numbers of products tested on animals and the amount spent on research on alternatives from 1984 through 1988. Greyhound returned its profile with that chart left blank.

Alternatives Research Program

Dial says it has been researching and attempting to validate a replacement to the Draize rabbit eye irritation test at an in-house in vitro program that it established in 1980, which the company says "has gained worldwide recognition." Company toxicologists have published several technical articles on a new method, which uses tissue cultures of rabbit corneal cells, thus eliminating the use of live animals.

The in vitro program on corneal cells is known as the SIRC cell colony forming assay. The company wrote IRRC that "These SIRC cells are exposed to different concentrations of test chemicals such as laundry detergent and the concentration which inhibits 50 percent of cell colony formation is identified. In most cases, the degree of cell toxicity in vitro has correlated well with the degree of eye irritation that might be found in animals, including humans."

Dial added: "For the past few years Dial's SIRC cell assay has also been a part of the Soap and Detergent Association and the Cosmetic, Toiletry and Fragrance Association found robin in vitro programs. These programs have been designed to involve several promising nonanimal tests in the evaluation of the potential eye irritation of household and laundry cleaning products and cosmetics. At this point, none of these tests has been validated or accepted by regulatory agencies, such as the Consumer Product Safety Commission and the Food and Drug Administration, in lieu of animal safety data. However, the agencies have demonstrated interest in the development of nonanimal tests and a willingness to accept in vitro data and evaluate it with respect to the corresponding in vivo data. The company continues to pursue approval of its nonanimal test program by regulatory agencies through the efforts of national cosmetic and laundry product trade associations."

Dial says it has contributed to several organizations that are working on alternatives. It says it helps to sponsor the Center for Alternatives to Animal Testing at Johns Hopkins University through the Cosmetic, Toiletry and Fragrance Association and helps CAAT review grants.

JOHNSON & JOHNSON

Overview of Johnson & Johnson's operations and animal testing

Breakdown of product lines	Number of animals used in 1988*	Percent of all animals used in 1988	1988 sales (millions)	Percent 1988 sales
Health care	NP	NP	$3,692	41%
Consumer Products	NP	NP	$5,308	59%
Totals	88,383*	100%	$9,000	100%

Source: USDA and Johnson & Johnson

*   Includes 67,410 rats and mice voluntarily reported by the company.
NP Not Provided

The majority of Johnson & Johnson's animal testing is done for its health care products—the pharmaceutical and professional medical products that made up 59 percent of the company's sales in 1988. The company classifies 40 percent of its business as being consumer products, but because the company did not break down its animal use by product line, it is impossible to determine how many animals were used to test the nonmedical consumer products that are the focus of this report. In a letter to IRRC, Johnson & Johnson explained, "Mergers and acquisitions confuse those usages, and some of our research organizations sporadically have reported rats and mice, so we don't have a useful, meaningful estimate for animals used for specific product categories." The company told IRRC that it did not use any outside laboratories "for the domestic testing of toiletries and nonmedical personal care products in 1988."

The company reported in spring 1989 that laboratory animal use for nonmedical consumer products has dropped 80 percent since 1983, "including animals used in eye and skin irritation studies," but it did not provide statistics to substantiate this figure. It appears from Johnson & Johnson's reports to the USDA that the company's total in-house animal use has dropped gradually in the last few years. Not including rats and mice, the company reported using 22,428 animals during 1987, compared with 25,453 in 1986, a decrease of 3,025 animals; in 1988, animal use again decreased, to 20,973. The information IRRC obtained in 1987 from the USDA does not include data on rats and mice; data for 1988 reveal that 76 percent of the animals used by the company in 1988 were rats and mice.

During 1988, Johnson & Johnson used a total of 39,993 animals in painful tests, although from the USDA reports it is clear that most of these tests were for new drugs. Seventy-five of the animals were guinea pigs used in a "sensitization experiment" by the Ethicon Research Foundation. In that case, the company said no pain relievers were administered because this could "have masked an acute toxic response of the parenterally administered material. All moribund animals were euthanatized immediately." The 1988 painful use total also includes 2,262 guinea pigs, 7,560 rats and 30,090 mice used by J&J's McNeil Pharmaceutical subsidiary; the company stated that painkillers would either "compromise the safety evaluation of new drugs being tested" or "invalidate the tests" performed.

Animals used at Johnson & Johnson

	1984	1985	1986	1987	1988
Dogs				3,239	2,711
Primates				149	35
Rabbits				7,595	6,902
Hamsters		Not		1,379	1,422
Guinea Pigs		Provided		10,175	9,690
Rats				NP	22,862
Mice				NP	44,548
Other				0	213*
Total without rats & mice			25,453	22,535	20,973
Total with rats & mice					88,383

Source: USDA and Johnson & Johnson

* 18 swine, 150 goats, 4 horses and 41 sheep

Animal Testing Policy

In outlining its approach to animal testing, the company says it is committed to "the three 'R' principles—replacement, reduction and refinement." It states that it supports "substituting non-animal systems, including cell cultures and lower organisms, for live animals,...using the fewest animals possible...[and using] techniques that limit the potential for discomfort to the animals."

The company told IRRC, "None of our toiletries or nonmedical personal care products contains any caustic or poisonous substance. Because we produce and test the mildest, gentlest products possible," the company said, animals experience "minimal or no pain whatsoever. The animals are not held in stocks and there is no need to use anesthetics during testing."

Noting that its research investment has gone up more than 60 percent since 1983, the company points out that it reduced the number of animals it uses in the same period. It cites protocols that use fewer animals, and the use of databases and other information sources, as ways in which the company has reduced its animal usage. "Additional reductions in the use of animals will depend a great deal on scientific validation and regulatory agreement," the 1989 proxy statement said.

Non-animal alternatives are now substituted for testing that used to be done on animals, the company reports. It says the methods it uses are being shared "with the scientific community through publications and presentations at symposia."

The company has established a Corporate Office of Animal Care and Use "to assure that the company's high standards are adhered to worldwide." Animal welfare at its facilities is well provided for "by conscientious and highly trained animal technicians/technologists and veterinarians," the company maintains.

In preparing draft profiles for this report, IRRC included a chart in which each company could show the numbers of products tested on animals and the amount spent on research on alternatives from 1984 through 1988. Johnson & Johnson returned its profile with that chart left blank. It said, "We cannot provide these data because we do not collect it in this manner."

Alternatives Research Program

The company says it continues to "search for additional procedures and actively pursue their validation, acceptance and adoption." To this end, Johnson & Johnson has provided financial support for its own and others' research on alternatives, the company reports. Johnson & Johnson told IRRC that it did not have exact records of its expenditures on alternative tests. "Several of our companies have independently initiated programs over the years to develop alternatives; some have failed, some look promising, but we have never tried to estimate the cost of these diverse efforts," the company said.

The Alternatives Report, published by the Center for Animals and Public Policy at Tufts University, noted in late summer of 1989 that J & J gave a three-year, $428,000 grant to the University of Texas at Austin, to "develop a reliable in vitro toxicity test for chemicals present in consumer products." The news-letter noted that in its first year, scientists in the program will culture cells from rabbit eyes to compare in vitro responses of the cells to chemicals that have already been tested on live rabbits. Also, rat cells will be used "as an experi-mental model for dermal toxicity." These cell cultures will be used as the basis for "in-depth" studies on cell toxicity in the program's second and third years. Dr. Daniel Acosta, the project's director, told IRRC that representative samples from different classes of chemicals will be used to test the sensitivity of the cell culture systems. So far, Acosta said, other scientists have used eye cells in experiments, but no "definitive study" has been produced yet. This program is among the first of J & J's outside grants for basic research on in vitro toxicology.

PROCTER & GAMBLE

Overview of Procter & Gamble's operations and animal testing

Breakdown of product lines	Number of animals used in 1988*	Percent of all animals used in 1988	1988 sales (millions)	Percent 1988 sales
Pharmaceuticals	NP	15%	NP	NP
Health care	NP		NP	NP
Food, other products	NP	85%	$4,495	23%
Totals	6,767*	100%	$19,336	100%

Source: USDA and Procter & Gamble

* 2,055 animals reported to the USDA by the company's Norwich Eaton Pharmaceutical subsidiary and 4,712 animals reported by Procter & Gamble's Cincinnati laboratories, where the company tests a combination of its health care, consumer and food products

NP Not Provided

Procter & Gamble conducts its animal testing at three company sites, and it contracts for additional testing with a number of contract laboratories. A company spokesman told IRRC in late 1989 that 15 percent of P&G's total animal use is for product safety testing of nonmedical consumer products. An independent assessment of the company's animal use figures by product line is not reachable from data reported to the USDA because the company tests a combination of its different product lines at its main laboratory in Cincinnati, and it uses outside laboratories.

In a letter to IRRC, the company said that it was "concerned about the heavy reliance on animal usage information" in IRRC's profile. "It has been our experience that discussions and comparisons focused on number are not productive and frequently confusing," P&G wrote. "For example, we expect to continue reductions in specific areas, but it is likely our overall numbers will level off or possibly even increase. There are several pressures affecting this, including new product introductions largely in the health care or drug area, new company acquisitions...[and] the fact [that] we believe we're close to the lowest level of animal usage we can obtain without concern for risking consumer safety until valid alternatives are developed and accepted."

P&G manufactures a wide range of consumer products. Personal care items include Ivory soap, Crest toothpaste, Scope mouthwash, Pampers disposable diapers, Sure deodorant and Charmin toilet paper; laundry and cleaning products made by P&G include Tide detergent, Comet cleanser and Downy fabric softener. Other products made by P&G are Folger's coffee, Crisco oil and Jif peanut butter, and such miscellaneous products as cellulose pulp and chemicals.

Painful tests reported to the USDA by Procter & Gamble's main laboratory and its leased facility in Cincinnati have gone down substantially in the last few years. In 1986, 31 dogs, 382 guinea pigs, 543 rabbits and two minipigs— a total of 958 animals—were used in painful tests without anesthesia. In 1987, the number fell to 152: three dogs, 53 rabbits and 96 guinea pigs. For 1988, 8 dogs, 14 rabbits and 54 guinea pigs, a total of 76, underwent painful tests without anesthesia.

The company reported to IRRC that 15 percent of its total animal use was for testing "non-food and non-drug products," while 85 percent was for the "development of regulated food and drug products, for which the government requires specific types of safety data." P&G says the total number of animals it used for all purposes at its own and outside laboratories declined 33 percent from 1984 to 1988. Consumer product safety studies used 83 percent fewer animals in 1988 than in 1984, the company reported. Based on the USDA reports of 1988 animal use at 6,767, and P&G's statements to IRRC that 85 to 90 percent of its animals are mice and rats that are not reported, and that less than 25 percent of overall use is through outside contractors that are not reported, an estimate would place P&G total animal use at no fewer than 60,000 and possibly more than 80,000 a year.

Animal Testing Policy

P&G defends the use of animal testing to ensure product safety because "animal testing in conjunction with the appropriate in vitro techniques remains the most scientifically acceptable approach." A P&G official told IRRC that at present some animal testing is necessary to provide safety assurance to federal regulators. The official said, "We've been working with the FDA and EPA to attain acceptance of alternatives to current animal testing procedures." However, the company's policy on animal testing commits the P&G to "the use of animals in testing only when necessary and only when no acceptable alternatives exist." It aims to "design testing programs so as to provide for humane treatment of all test animals and minimize numbers of animals used."

In response to questions about the company's animal testing at its October 1989 annual meeting, the company's president, John Smale, said, "We do not approach this issue lightly. Animal tests are expensive, they are time-consuming, and they are difficult to administer. They also subject us to letter writing campaigns, and on-going explanations and debates." He added, "On the other hand, our clear responsibility to assure that our products are safe for human use mandates that we run some tests with animals."

"It is true that federal laws do not mandate the use of animals in confirming product safety. But it is equally true that the government regulators who are charged with interpreting and enforcing these laws continue to require these tests," Smale said.

Decreases have come from a number of changes in methodology, P&G says. The company uses a database of safety test results to avoid repeating relevant tests. By predicting the toxicity of a substance through computer-based models, the company reports, it has been able to cut back further on its animal use. In late 1989, P&G told IRRC it "expects to be able to cut back further in some areas. These reductions have occurred at a time when the company's overall business has been expanding, especially in the health care area. In the last seven years, P&G has acquired three major health care businesses."

The company uses an oral toxicity test that it says requires only one-fourth the number of animals used by the classical LD50. The company also cites its use of a low volume eye irritation test that subjects rabbits to one-tenth the volume of test substance required in the traditional Draize eye irritation method. After experiments with both humans and rabbits, the company reported in 1986 that the low volume test more accurately predicted human responses and caused less stress for the rabbits than the traditional Draize test, which P&G does not use.

In addition, the company says it significantly reduced animal use by substituting four in vitro tests for the first tier of a method that formerly used 600 animals to determine a chemical's potential for causing cancer or mutations.

In his statement at the 1989 annual meeting, Smale also said, "We are aware that some cosmetics companies have announced moratoriums or the elimination of animal testing. I can't judge individual company actions. I can tell you, however, that one of these firms has acknowledged publicly that, 'suppliers to the cosmetics industry may have to continue to do some animal testing to substantiate the safety of their new ingredients.' These companies surely recognize current scientific limitations," Smale concluded.

**Animals used at Procter & Gamble's Cincinnati
facilities for product testing***

	1984	1985	1986	1987	1988
Dogs	294	186	474	415	399
Cats	0	2	6	0	0
Rabbits	2,118	1,486	1,430	685	236
Hamsters	581	522	45	263	845
Guinea Pigs	7,789	4,740	4,217	3,248	2,886
Other**	0	0	684	189	346
Totals	10,782	6,936	6,856	4,800	4,712
No. of products tested***	72	77	107	107	117

Source: USDA and Procter & Gamble

* This table includes figures reported to the USDA for 1988 from Procter & Gamble's main laboratory and leased facility in Cincinnati; while some consumer products were tested at these facilities, a number of the company's other products (health care and food items) were also evaluated.

** The 684 "other" animals in 1986 were 605 gerbils, 65 ferrets and 14 minipigs; IRRC does not have species data on the "other" animals used in 1987; in 1988, the "other" animals used were all ferrets.

*** P&G told IRRC, "This product count considers our hair care product lines (e.g., Vidal Sassoon, Pantene), baking mixes and skin care lines as a single product. The many variations of the brands are *not* included in this count."

Alternatives Research Program

P&G cites several new programs that may help supplant the use of animals in product research, including:

- An in vitro assay program to evaluate four promising tests as alternatives to ocular irritancy testing. The assays included are: Microtox; Light addressable Potentiometric Sensor; Neutral Red and Tetrahymena Thermophila Motility.

- Using tissue culture and organ culture systems to detect a substance's potential to cause birth defects;

- Adapting cell technology to evaluate immune system and allergy responses;

- Analyzing animal responses with non-invasive techniques, such as nuclear magnetic resonance imaging; and

- Measuring changes in organ function, which already has decreased the number of animals used.

A P&G official told IRRC that the company spent approximately $3.9 million during 1989 on research devoted to the development of testing alternatives, and more than $10 million during the last three years. P&G also contributes to the Cosmetic, Toiletry and Fragrance Association's efforts to find alternatives, and it supports the Soap and Detergent Association's project to identify alternatives to eye irritation tests. The company policy, according to its 1988 proxy statement, "is to publish the results of its alternative test method research so that the learning can be available to everyone. A P&G official told IRRC in late 1989 that its scientists have published more than 70 papers on alternatives since 1987, "helping to advance scientific and regulatory acceptance of alternative methods. In June 1989, P&G announced a new University Animal Alternatives Program that provides grants totaling up to $150,000 annually and $450,000 annually after the third year. This represents additional new funding for alternatives research. The company expects to award its first grants in early 1990."

Henry Spira, a prominent animal activist who pressured Revlon and other cosmetics companies to finance research on alternatives to animal testing, turned to Procter & Gamble and other firms to lead the way in changing corporate policies toward animal testing—especially concerning the LD50. In the 1983 report of the Coalition to Abolish the LD50, Spira praised P&G for its "thoughtful, detailed and integrated" new toxicology program. He wrote in a letter that year to P&G that its "initiatives and commitment to replace and reduce the use and suffering of lab animals [is] both visionary and practical." In 1986, *The Wall Street Journal* focused on P&G in a report on new laboratory techniques and policies. In 1988, the Council on Economic Priorities, a consumer advocacy group, presented Procter & Gamble with a Corporate Conscience Award in the Animal Rights category. According to the company's 1988 proxy statement, this award "recognizes the company's longtime leadership in seeking alternative testing for animals." In 1989, the Massachusetts Society for the Prevention of Cruelty to Animals gave a similar award to P&G.

The company's scientists have made various presentations to leading scientific and regulatory bodies and have served on panels and task forces concerning animal use for the congressional Office of Technology Assessment, the European Chemical Industry Ecology and Toxicology Center, and the New York Academy of Sciences.

Controversy over P&G Public Relations Plan

On July 26, 1989, the *Cincinnati Enquirer* published a story about an initiative planned by P&G that would "promote the judicious use of animal testing." Animal protection activists learned about the plan from an internal company memo they obtained. Activists said the memo revealed that the company was not really committed to ending animal use and would divert funds that would be better used for alternatives research. In Defense of Animals, an animal rights group, announced a boycott of P&G and organized protests around the country.

The memo stated that the mission of the planned industry coalition was "To sustain a public and legislative environment which supports the judicious use of animal testing as a necessary part of a corporation's responsibility to manufacture and market safe and effective products." It said the group would be independent of but coordinated with similar activities by several other industry support groups. Proposed members of the coalition that had "already indicated interest" were Bristol-Myers Squibb, Johnson & Johnson, Merck and Syntex, in addition to P&G. Other potential members the memo listed were Eastman Kodak, IBM, Monsanto, Minnesota Mining & Manufacturing, Colgate-Palmolive, Lever Brothers and Gillette. In all, the proposed budget was $17.5 million for the first three years of the coalition.

P&G's president John Smale, in his statement at the 1989 annual meeting, commented on the controversy, which he characterized as a "misunderstanding growing out of circulation of a stolen internal P&G memo proposing the formation of an industry group to sponsor a public education program on this issue." He said that if the program proceeded, its objective would be "simply to help consumers understand what is now possible and what isn't. We think Procter & Gamble and other companies working on development of alternatives to animal testing have an obligation to keep the public informed. If such a program materializes, its cost to Procter & Gamble would be but a small fraction of the investment we are making in finding alternatives to animal testing."

SCHERING-PLOUGH

Overview of Schering-Plough's operations and animal testing

Breakdown of product lines	Number of animals used in 1988*	Percent of all animals used in 1988	1988 sales (millions)	Percent 1988 sales
Pharmaceuticals	NP	NP	$2,210	74%
Consumer Products	NP	NP	$760	26%
Totals	60,717*	100%	$2,969	100%

Source: USDA and Schering-Plough

* Includes 39,348 rats and mice
NP Not Provided

The vast majority of animals used for testing at Schering-Plough are for the company's pharmaceutical and health care business. Animal testing also occurs for Schering-Plough's consumer product lines, which include Maybelline cosmetics, Coppertone and other sun protection products. On Dec. 12, 1989, Schering-Plough announced plans to sell its Maybelline cosmetics business, because of the company's intention to concentrate on pharmaceuticals, "personal care and vision care products." The company told IRRC that it used no outside contractors for Maybelline animal testing in 1987 or 1988, but it did not indicate the extent of contract testing for its pharmaceuticals or other consumer products.

In 1988, IRRC obtained what may have been Schering-Plough's first public disclosure of the number of animals, by species, used in testing Maybelline products from 1982 to 1987. The company declined to tell IRRC how many animals were used for testing its Maybelline cosmetic formulations in 1988, however, and also has not provided information on testing for its other consumer products. An official did indicate to IRRC in spring 1989 that the company's animal testing program has not changed significantly since 1987.

Regarding requests for data on painful experiments, the company told IRRC, "In the opinion of both our in-house scientists and our outside reviewing experts, none of these tests involves pain and our annual reports to the Department of Agriculture so state."

The figures on Maybelline's animal usage by species that Schering-Plough gave IRRC in 1988 covered the period 1982-87, and included a tabulation of new and revised Maybelline formulations (i.e., products) tested for those years. In 1982, the company tested 84 percent of its cosmetic products on 808 animals, while in 1984, the percentage of products tested on animals had fallen to 71 percent, and 674 animals were used. In all, for the years reported, Maybelline reduced total animal use in four of five years (1986 was an exception). The 1986 increase was caused by "totally new cosmetic formulations with unique components," the company said.

Animals used at Maybelline

	1984	1985	1986	1987	1988
Rabbits	360	324	518	306	NP
Guinea Pigs	118	120	120	0	NP
Rats	100	50	70	150	NP
Totals	578	494	708	456	NP

Source: Schering-Plough

Animal Testing Policy

In 1988, a company official acknowledged to IRRC, "The federal agencies do not say 'Thou shalt test thy products on animals.' What they say is that you should do adequate safety testing." However, the official asserted that "it is our interpretation of the law that, in the case of the animal testing that we do, there is no adequate non-human way of testing." In a position statement provided to IRRC in late 1989, the company added that it is "dedicated to maintaining the highest standards of humane laboratory animal care and to utilizing the most appropriate in vitro and in vivo [animal] models for evaluating the efficacy and safety of new pharmaceuticals."

A company official told IRRC that its reduction in animal use is "largely as a result of our and our suppliers building up a base of toxicological knowledge over the years. This allows us to make an increasing number of safety judgments without the need for repetitive testing." Indeed, the company's data show that at least until 1987 it consistently cut down the proportion of formulations that were tested on animals, from 84 percent in 1982 to 31 percent in 1987 (see chart below for data on the last five years).

The company also says that it has modified the controversial Draize test by using six rather than nine rabbits, and that six animals are used for most skin and eye irritancy tests (some skin tests use 25 guinea pigs). The company told IRRC in 1988 that "no LD50 tests are performed and no lethal dosages are sought." As a modification of the classical LD50 method, 10 animals are used for acute oral toxicity tests. Instead, Maybelline performs safety tests on animals "to assure non-toxicity in the case of accidental ingestion" by humans. A company official explained that the company "tests a moderately large dose slightly in excess of the amount that a two year old child could ingest by swallowing the whole product." It says the only other animal tests conducted by Maybelline involved such skin responses as irritancy and allergenicity.

Animals are often used more than once by the company; notably, the six rabbits in a skin test are always the same rabbits used for the eye tests. Also, a skin test may involve several formulations. Such multiple testing has kept the ratio of total animal use to tested formulations well below the least number of animals—six—tested on a given formulation.

Maybelline products tested with and without animals

	1984	1985	1986	1987	1988
No. of products tested on animals	174	164	231	141	NP
No. of products tested w/o animals	167	238	389	309	NP
Total number of products	341	402	620	450	NP
Percent of products tested on animals	51%	41%	37%	31%	NP

Source: Schering-Plough

Schering-Plough also reports that it supported "the development of guidelines for the ethical care and use of laboratory animals through the New York Academy of Sciences Ad Hoc Committee on Animal Research."

The company declined to provide information on the amount of money spent on either animal or nonanimal tests.

Alternatives Research Program

In its 1989 proxy statement, the company reported that it is "evaluating the agarose diffusion method, which has been recently shown to be potentially useful as a preclinical screen for product safety tests. This test, if validated, could diminish further our use of animals in cosmetics testing." The company declined to provide further information on its research into alternatives when asked by IRRC.

The company told IRRC that it has supported research at the Center for Alternatives to Animal Testing at Johns Hopkins since its inception, and has supported the Rockefeller University Laboratory for In Vitro Toxicology Assay Development. The company also says it has been involved in trade association activities to explore alternatives to animal testing. Schering-Plough did not provide information on how much money it has contributed to these efforts.

V: ANALYSIS OF ANIMAL USE REPORTED TO THE USDA

An impeccable repository of statistics on animal use trends in the United States does not exist. As concern over the use of animals in research and testing has grown, different observers have made varying estimates. While animal protection activists have at times said that as many as 100 million laboratory animals are used annually in this country, other estimates put the figure closer to 10 million. In its 1986 report, the congressional Office of Technology Assessment projected an annual use figure of about 22 million. No truly accurate figures exist, however, because there are no requirements for all facilities that use animals to report all species to any central agency.

The OTA report found that the most reliable source of animal use data was the U.S. Department of Agriculture's Animal and Plant Health Inspection Service (APHIS). The Animal Welfare Act requires facilities that use certain species of animals in research and testing to file annual reports with APHIS. The reports, submitted at the end of each fiscal year, divide animal use into three categories: nonpainful procedures, painful procedures in which animals receive anesthesia and painful procedures in which no relief is provided. Dogs, cats, rabbits, primates, guinea pigs, hamsters and wild animals must be reported. Reports on mice and rats, which account for an estimated 80 to 90 percent of all animal use, are not required, although some researchers voluntarily report their use of these species.

The annual reports to APHIS are stored on a computer database, which is used to generate an annual report to Congress on animal use by state and category of experiment, as well as on the agency's enforcement activities and record of inspecting animal laboratories during the past year.

Companies Reporting the Highest Number of Animals Used
In Research and Testing in Fiscal Year 1988

Facility	Type of Test			
	Not Painful	Painful with Anesthetic	Painful with No Anesthetic*	Total
SmithKline Beckman	35789	786	0	36575
Imperial Chemical Industries#	14840	15688	4092	34852
Merck	5201	17908	203	23312
Wellcome#	7638	290	15268	23196
Bayer#	11228	381	10800	22409
Rorer Group	18613	3462	0	22075
Eastman Kodak	7810	13799	322	21931
Schering-Plough	19344	204	1835	21383
Johnson & Johnson	8273	10357	2343	20973
American Home Products	17012	3851	96	20959
American Cyanamid	11746	1733	5362	18841
Eli Lilly	9654	4335	4794	18783
Hoffmann-La Roche#	13316	5263	0	18579
Pfizer	17382	760	256	18398
Boehringer-Ingelheim#	8856	1358	7307	17521
Du Pont	6877	8327	730	15939
Fermenta#	14553	0	0	14553
Corning Glass Works	13326	644	186	14156
Southern Research Institute	13323	50	0	13373
Warner-Lambert	11655	375	1207	13237
Monsanto	3960	7402	468	13035
South Mountain Laboratories	12682	87	0	12769
Baxter International	9534	1730	168	11432
Hill Top BioLabs	10070	0	960	11030
Green Cross Corp.#	10157	40-	0	10197
Upjohn	7403	2490	126	10019
Leberco Testing	8607	32	1328	9967
Beecham Holdings	8954	0	0	8954
North American Science Associates	7508	1434	0	8942
Diamond Scientific	4193	522	4188	8903
Polysar Energy & Chemical#	8792	0	0	8792
Abbott Laboratories	7301	1071	0	8372
Dow Chemical	6156	1229	226	7611
Internat'l Research & Development	7107	33	0	7140
Syntex	4570	1793	646	7009
Nestle#	4641	2287	38	6966
Procter & Gamble	4981	1698	112	6791
A.H. Robins	891	693	5207	6791
CIBA-GEIGY#	5349	893	496	6738
Northview Pacific Labs	3927	221	2345	6493
Biosearch	5420	188	748	6356
Ent. de Recherch. & d'Act. Petr.#	6278	0	0	6278
Minnesota Mining & Manufacturing	2678	2333	1110	6121
Squibb & Sons##	2512	3404	0	6112
Solvay & Cie#	5747	300	0	6047
Colgate-Palmolive**	5041	999	2	6042
Bristol-Myers##	2152	3405	169	5726
Consumer Product Testing	4835	743	0	5578
Applied Bioscience International	4932	147	355	5434
International Minerals & Chemical	5384	7	0	5391
Totals:	448198	124752	73493	648081

* The U.S. Department of Agriculture reporting form states: "Number of animals used in research, experiments, or tests involving pain or distress without administration of appropriate anesthetic, analgesic, or tranquilizer drugs."
# Foreign-owned facility
##Bristol-Myers merged with Squibb in 1989, forming Bristol-Myers Squibb
**Includes Colgate-Palmolive's Kendal McGaw, which was sold in Oct. 1988

Source: USDA data compiled by IRRC

Noncommercial Facilities Reporting the Highest Number of Animals Used
In Research and Testing in Fiscal Year 1988

| Facility | Type of Test | | | |
	Not Painful	Painful with Anesthetic	Painful with No Anesthetic*	Total
University of California	33020	39179	1097	73296
University of Wisconsin	43896	4028	0	47924
University of Georgia	12164	6176	20	18360
Johns Hopkins University	1769	13754	0	15523
Univ. of Texas Southwestern Med Ctr	7753	6294	295	14357
University of Minnesota	7507	4216	10	11934
University of Iowa	3919	7059	0	10978
Emory University	6227	4634	6	10867
University of Michigan	4077	6606	32	10715
State University of New York	3688	6742	0	10590
University of Florida	5452	4806	141	10399
NYS Dept. of Health & Health Research	7662	1445	0	9107
Louisiana State University	2790	5847	0	8637
New York University	5222	2984	0	8206
Univ. of Illinois at Urbana-Champaign	1532	740	30	8040
Univ. of Texas Health Science Ctr	1587	6050	354	7991
University of Washington	2879	4628	2	7509
Washington University	2974	4462	0	7436
Baylor College of Medicine	911	6505	15	7431
University of Delaware	3913	3422	0	7335
University of North Carolina	2859	4304	24	7187
Harvard University Medical School	6042	955	0	6997
Ohio State University	3006	3932	0	6938
University of Illinois at Chicago	5048	1677	0	6725
University of Missouri	1914	1882	126	6585
University of Cincinnati	342	6075	132	6549
Indiana University	3563	2813	0	6376
Colorado State University	3224	3050	100	6374
Vanderbilt University	4954	1379	0	6333
Medical College of Georgia	22	6270	0	6292
University of Virginia	459	5800	8	6267
University of Louisville	2107	4068	0	6175
Duke University	479	5517	0	5996
Mayo Clinic/Foundation	1705	4267	0	5972
Yale University School of Medicine	3148	2561	224	5933
University of Arizona	1885	4000	0	5885
Univ. of Alabama at Birmingham	4198	1659	0	5857
Univ. of Texas Med Branch-Galveston	2672	2883	0	5555
Stanford University	3462	2068	0	5530
Massachusetts General Hospital	3327	2120	0	5447
Northwestern University	3357	2089	0	5446
Wake Forest University	2662	2690	0	5352
Virginia Commnwealth University	1634	3657	36	5327
Southwest Fndn for Biomedical Res.	1302	3857	0	5159
University of Massachusetts	2731	2368	0	5099
University of Maryland at Baltimore	2786	2136	36	4958
University of Pennsylvania	1913	2864	0	4777
University of Nebraska-Lincoln	2898	1875	0	4773
University of Southern California	1992	2606	0	4598
Case Western Reserve University	1277	3305	0	4582
Totals:	235910	234304	2688	481679

* The U.S. Department of Agriculture reporting form states: "Number of animals used in research, experiments, or tests involving pain or distress without administration of appropriate anesthetic, analgesic, or tranquilizer drugs."

Source: USDA data compiled by IRRC

To make a general assessment of how many animals were used for what purposes, IRRC made a series of Freedom of Information Act requests to APHIS over the last three years. IRRC first obtained copies of the computer databases that had information from annual research reports for fiscal years 1986 and 1987. These databases did not include reports submitted after Sept. 30 of each year, and IRRC was unable to determine the degree of error caused by data entry errors of USDA staff, although clearly some mistakes had been made. Therefore, in requesting 1988 data, to avoid possible problems caused by electronic data transmission, in fall 1989 IRRC obtained paper copies of all the annual reports filed for fiscal year 1988 and constructed its own database, similar to that it had received for 1986-87.

For each year, IRRC categorized reporting facilities as commercial or noncommercial. Subsidiaries of companies listed in reference books on standard corporate affiliations were combined under the parent company's name. To determine what kinds of institutions were the heaviest users of animals, IRRC then ranked the two categories of users. Tables showing the 50 companies and noncommercial facilities that reported the most animals used for fiscal year 1988 appear on pp. 130 and 131. All figures in this chapter contain USDA data compiled by IRRC.

Limits to the USDA data: The USDA information for each company does not include animals used in tests conducted by the contract laboratories routinely used by many companies for some product testing; these laboratories are not required to report the allocation of animal usage among their corporate clients, so information on such testing is unavailable, except where a company voluntarily reveals it. Rats and mice also are not included in the totals, since reporting for these species is not consistent. These omissions present significant constraints to the USDA data. However, despite the consequent understating of corporate animal use, the data provide the most inclusive look at animal use now available.

One other flaw in the USDA's reporting procedure is that defining what constitutes a painful test is up to the reporting facility. Critics of the current reporting method say this means judgments about the pain experienced by research animals are completely subjective, and that the USDA's categories that separate painful tests from nonpainful tests have little meaning. Although this may be a valid criticism, the statistics do give at least some idea of how many animals experience pain in research and testing in the United States.

As discussed in the company profiles in Chapter IV, the USDA data generally are not specific enough to provide a clear picture of consumer product testing at companies, since firms are not required to report their animal use by product category, and many companies test a broad variety of items in health care as well as household products categories.

Findings from the USDA data: Overall, general trends in *how* animals are used did not seem to change dramatically from 1986 to 1988, although the *number* of animals reported used appears to have dropped by about 100,000 animals from 1986 to 1988, despite an uptick in 1987. In this period overall, facilities classified about 6 percent of their animal use as painful without anesthesia, about 35 percent as painful with some kind of anesthesia and the remaining 59 percent as not painful.

Reported animal use for all three years was split almost evenly between commercial and noncommercial users. In 1988, while companies reported about 49 percent of all animal use, colleges and universities used approximately 44 percent, hospitals and local government agencies reported around 5 percent and other noncommercial facilities used about 3 percent of all animals.

In fiscal year 1988, all commercial users reported a little more than 800,000 animals used. (While pharmaceutical and chemical companies report the most animal use, many of the companies that use the most animals also have contributed substantial time and funds to the search for alternatives to animal testing.)

Painful tests—Of particular concern to animal welfare proponents are painful tests. The most striking difference between commercial and noncommercial animal use was in the "painful without anesthesia" category. In 1986 and 1987, 90 percent of all such use was reported by corporations. In 1988, the figure dropped slightly, to 88 percent of all reported painful procedures. According to the USDA data, only about 1 percent of the animals used by noncommercial users experienced pain, while 11 percent of the animals used by companies fell into this category.

Eighty-eight percent of the approximately 97,000 painful procedures reported by all facilities in 1988 were conducted by companies. This figure is somewhat higher than the one reported by APHIS (90,400) to Congress, probably because the IRRC database includes reports submitted after Sept. 30, and the report to Congress does not. The 10 companies that reported the largest number of painful tests made up 72 percent of all such tests among corporations. Seventeen companies reported using more than 1,000 animals in painful tests, while five firms reported using more than 5,000 animals in painful procedures.

Noncommercial users such as schools and hospitals tend to use anesthesia much more often than their commercial counterparts because they use ani-

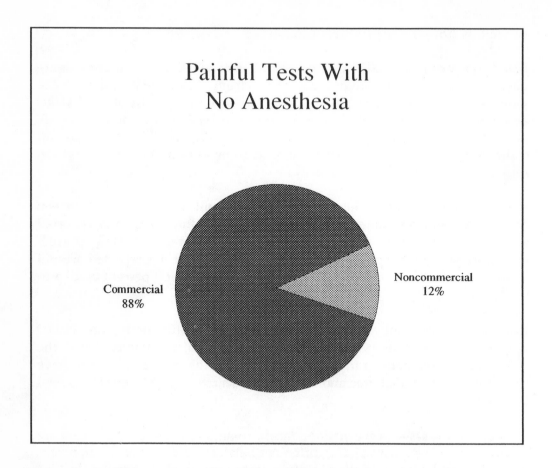

Painful Tests With No Anesthesia

Commercial
88%

Noncommercial
12%

mals differently. Companies conduct acute toxicity tests and skin and eye irritation studies to find out how their products may affect human users. Some of these tests are by definition painful. Further, vaccination products evaluated by companies also generally inflict pain on laboratory animals. Facilities that report painful use must provide a brief explanation for these procedures; researchers' comments on the USDA forms usually say that pain relievers would interfere with test results and skew the resulting toxicity data. Schools and hospitals, on the other hand, generally conduct basic biological, medical and educational research for which some form of anesthesia is usually considered acceptable to the experiment.

Differences in use of species—Corporations were most likely to use guinea pigs (40 percent), rabbits (29 percent) and hamsters (17 percent), while noncommercial facilities divided their use more evenly among rabbits (25 percent), guinea pigs (20 percent) and hamsters (11 percent). Reports of use of dogs, cats and primates—the species most likely to evoke protest when used—were less frequent. Cats and primates each made up 1 percent and dogs 6 percent of corporate animal use; for noncommercial use, dogs were 11

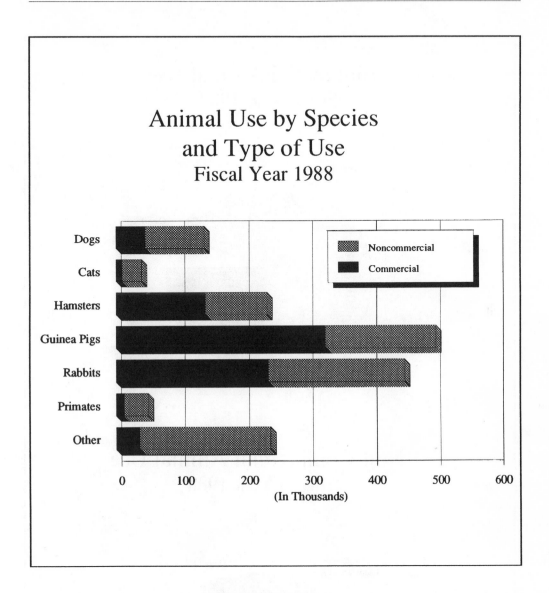

percent, primates 5 percent and cats 4 percent.

The USDA's "other animals" category, ostensibly only for wild animals and warm-blooded animals not otherwise counted, in practice tends to be a catch-all classification, with reported species ranging from cows and pigs to voles and raptors. Just 5 percent of animals in commercial use were classified as "other" species, but fully 15 percent of the animals reported by noncommercial users were put in this category.

IRRC's figures differ slightly from the data reported to Congress by APHIS for 1988; late reports not included in that report are probably the cause.

Commercial Animal Use
Fiscal Year 1988

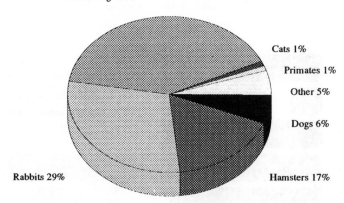

Guinea Pigs 40%

Cats 1%

Primates 1%

Other 5%

Dogs 6%

Rabbits 29%

Hamsters 17%

Non-Commercial Animal Use
Fiscal Year 1988

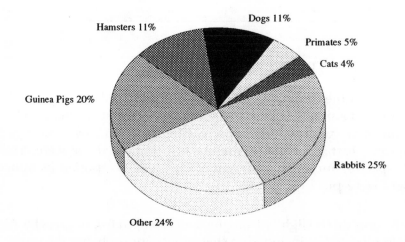

Dogs 11%

Hamsters 11%

Primates 5%

Cats 4%

Guinea Pigs 20%

Rabbits 25%

Other 24%

Concentration of animal use among top users—The 10 firms that used the most animals accounted for approximately 30 percent of all commercial use, and the top 50 firms' animal use made up around 80 percent. In contrast, animal use among noncommercial facilities was spread more evenly: Although the top 10 users accounted for about 28 percent of all noncommercial animal use, the top 50 facilities used only about 57 percent.

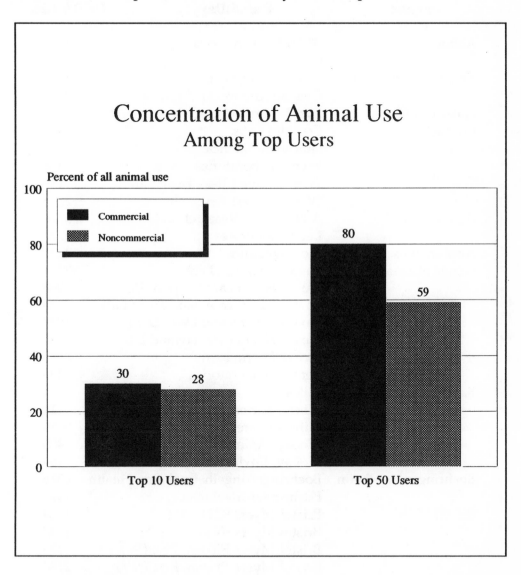

Parent-subsidiary relationships
in IRRC's list of the top 50 commercial animal users
(fiscal year 1988)

Parent	Subsidiary	USDA Lic.
Abbott	Abbott Laboratories	3325
	Oximetrix	93166
American Cyanamid	American Cyanamid	2293
	Cyanamid Fndn f/ Agri. Dev.	2210
American Home	Fort Dodge Laboratories	429
Products	Franklin Labs	7443
	Sherwood Medical	5814
	Wyeth Laboratories	2318
	Wyeth-Ayerst Res., Drug Safety	21139
	Wyeth-Ayerst Research	2295
A.H. Robins	A.H. Robins Research Labs	5210
	Lee Laboratories	5723
Applied Bioscience	Bio/Dynamics	2233
Baxter International	Baxter Edwards Labs	93121
	Baxter Healthcare Corp. of P.R.	948
	Baxter Healthcare, Bax. Tech. Park	3323
	Baxter Healthcare, Dade Div.	5828
	Baxter Healthcare, Hyland Div.	93268
	Baxter Pharmaseal	9363
	Bentley Laboratories	9393
Bayer	Miles	321
	Miles Pharmaceuticals	9111
	Miles—Cutter Biological	9318
	Mobay, Animal Health Division	485
	Mobay, Environmental Health Res.	484
Boehringer-Ingelheim	Boehringer-Ingelheim Anim. Health	4314
	Boehringer-Ingelheim	1629
Bristol-Myers	Bristol-Myers (CT)	1634
	Bristol-Myers (NY)	2148
	Bristol-Myers Pharm. Res. (IN)	325
	Bristol-Myers Pharm. Res. (NY)	2134
	Bristol-Myers Products	2221
Ciba-Geigy	Ciba-Geigy	1632
	Ciba-Geigy Pharmaceuticals	2214

Parent	Subsidiary	USDA Lic.
Colgate-Palmolive	Colgate-Palmolive	21164
	Kendall McGaw	9371
Corning Glass	Hazelton Labs America (VA)	526
Works	Hazelton Labs America (WI)	3530
Dow	Dow Chemical	348
	Merrell Dow Research Institute	3184
Du Pont	A.I. Dupont Institute	509
	Du Pont (DE)	502
	Du Pont (PA)	23108
	Du Pont Critical Care	3334
	Du Pont NEN Products	1423
	Haskell Laboratory	501
	Stine Laboratory	5010
Eastman Kodak	Eastman Kodak	2187
	NeoRx	9129
	NOVA Pharmaceutical	5142
	Sterling Drug	4824
	Sterling Research Group (NY)	2111
	Sterling Research Group (PA)	23138
Eli Lilly	Cardiac Pacemakers	4122
	Hybritech	93227
	Eli Lilly	323
Fermenta	Fermenta Animal Health (CO)	8436
	Fermenta Animal Health (NE)	475
	Ricerca	3191
Johnson & Johnson	Ethicon Research Foundation	2212
	J&J Baby Products	2287
	J&J Biotechnology Center	93202
	J&J Research Foundation	2216
	McNeil Pharmaceutical	2315
	Ortho Diagnostic Systems	2264
	Ortho Pharmaceutical	226
Int'l Minerals &	Mallinckrodt	435
Chemical	Pitman-Moore (IN)	3223
	Pitman-Moore (NJ)	2244
Monsanto	G.D. Searle	3328
	Monsanto	4333
Nestle	Alcon Laboratories	7411
Pfizer	Valleylab	8431
	Pfizer	2188
Procter & Gamble	Norwich Eaton Pharmaceuticals	2122
	Procter & Gamble	3110

Parent	Subsidiary	USDA Lic.
Rorer	Armour Pharmaceutical	3333
	Rorer Central Research	239
Sandoz	Sandoz Research Institute	229
	Zoecon	7435
Schering-Plough	DNAX Research Institute	93221
	Plough	6310
	Schering (NE)	4710
	Schering (NJ)	2236
SmithKline Beckman	Allergan	9367
	Beckman Instruments	9374
	Norden Laboratories	4714
	SmithKline Beckman	2312
Solvay & Cie	Fromm Laboratories	3512
	Salisbury Laboratories	4212

Appendix A: Directory of U.S. Animal Protection Groups

Activists in the animal protection movement maintain that more than 10 million people are involved in its activities, and one estimate puts the number of animal rights and animal welfare groups nationwide at 7,000. Following is a sampling of those groups that have a particular interest in animal testing. In developing this list, IRRC started with a compilation in the *Animals' Voice* magazine and added other groups that turned up in literature on the testing debate. We then sent a short questionnaire to all of the organizations. What follows is based on responses to the questionnaire. While not comprehensive, the list gives a good indication of the variety and nature of groups that oppose animal testing.

While the groups listed here by no means constitute a complete listing of all organizations in the animal rights movement, they are representative of the movement. Of the 47 respondents to IRRC's questionnaire, the 35 that gave membership information reported a combined total of 2.8 million members. The smallest group that provided this information—the Animal Legal Defense Fund—had 250 members, while the largest—The Humane Society of the United States—reported 400,000 members. Only 29 respondents gave information about their budgets. Of these, the total reported came to $49 million, ranging from $15,000 for the Humane Education Committee to almost $16 million for the Massachusetts Society for the Prevention of Cruelty to Animals. Many of the organizations were founded in the late 1970s and early 1980s; only three were founded before 1950.

THE ALLIANCE FOR ANIMALS
P.O. Box 909
Boston, MA 02103
Telephone: (617) 265-7577

Number of Members: more than 2,000
Year Founded: 1988
Annual Budget: $100,000
Cost to Join: $15

The organization describes itself as follows:

> We mainly work on the fur issue and local care for stray animals, but legislative and educational work has concerned animal testing. Members canvassed door to door with information on cosmetics testing when a bill was being considered in the Massachusetts legislature. The group sees itself as a liaison between legislators and scientists concerned with animal testing. Also, the group provides information on cruelty-free products and where they can be found.

AMERICAN ANTI-VIVISECTION SOCIETY
Suite 204 Noble Plaza
801 Old York Road
Jenkintown, PA 19046-1685
Telephone: 215-887-0816

Number of Members: not provided
Year Founded: 1883
Annual Budget: not provided
Cost to Join: $10

The organization describes itself as follows:

The American Anti-Vivisection Society is a national organization dedicated to the abolition of animal use in research, testing and education. We are motivated by the conviction that the use of animals in the laboratory is morally wrong, and that animals have a right to be free from such exploitation. In addition to that ethical concern, our programs, campaigns and literature reflect our belief that animal experiments are characterized by fundamental scientific limitations, chief among them the difficulty of transferring results to human beings with any degree of reliability....Among our current interests are state-level initiatives to ban product testing with animals.

AMERICAN FUND FOR ALTERNATIVES TO ANIMAL RESEARCH (AFAAR)
175 West 12th Street, #16G
New York, NY 10011-8275
Telephone: (212) 989-8073

Number of Members: about 7,000
Year Founded: 1977
Annual Budget: $40,000 to $200,000
Cost to Join: $10 plus

The organization describes itself as follows:

> AFAAR raises funds from the public in order to offer scientific grants to develop nonanimal methods of testing, research and education.

Since its founding, the group has

provided grants totaling more than $600,000 to fund research into alternative testing methods.

The group says, "The main hope of any sizable reduction in the use of laboratory animals, including very painful uses, is cooperation between animal protectionists and biologists to develop, evaluate and use alternatives....We are convinced that goal-directed research for these substitutes is legitimate." AFAAR believes that "industry and government departments which use animals" should provide the main support for alternatives, because "funds available from animal welfare are limited, while the needs of all branches of animal welfare are great." The group concludes, "Our correspondence leads us to believe that a well-informed public will not countenance continued use of animals without at least a determined attempt to replace them."

AMERICAN SOCIETY FOR THE PREVENTION OF CRUELTY TO ANIMALS

441 East 92nd Street
New York, NY 10128
Telephone: (212) 876-7700
Fax: (212) 348-3031

Number of Members: 350,000
Year Founded: 1866
Annual Budget: $10,871,000
Cost to Join: $20

The organization describes itself as follows:

The ASPCA, the first humane society in America and, today, one of the largest humane societies in the world, was incorporated in 1866 by a special act of the New York State legislature. The ASPCA's purpose, as stated by its founder, Henry Bergh, is "to provide effective means for the prevention of cruelty to animals throughout the United States." The society encourages reverence for life and respect for all animals as sentient creatures deserving of humane care and protection. Its work encompasses (1) companion animals, (2) animals in research and testing, (3) animals raised for food, (4) wild animals, (5) entertainment and work animals, and (6) animals in education.

Specifically in regard to animals in research and testing, the society believes that unwanted animals from pounds or shelters should not be used for research or testing. Further,

The use of animals for experimentation should be permitted only when there are no known feasible alternatives and then only when the experiment is believed likely to produce new and substantive information. In such instances the experiments must be carefully designed to use the smallest number of the most suitable species, maintained in a sanitary environment and humanely cared for before, during and after any experimental procedure. Every effort must be made to eliminate pain and suffering.

Scientists should give priority to the development of alternatives to animal-based research, with the ultimate goal being to eliminate the need for any research on animals.

ANIMAL ALLIES
P.O. Box 35063
Los Angeles, CA 90035
Telephone: (213) 936-5166

Number of Members: 10,000
Year Founded: 1985
Annual Budget: varies—depending upon emphasis on each campaign
Cost to Join: $10

The organization describes itself as follows:

Goal: End animal abuse and exploitation wherever it occurs. Projects: antivivisection campaigns, pound seizures campaign, buck rodeo campaign, end pet stores and puppy mills, end factory farming, support veganism, anti-fur campaign.

ANIMAL LEGAL DEFENSE FUND
1363 Lincoln Avenue
San Rafael, CA 94901
Telephone: (415) 459-0885

Number of Members: 250
Year Founded: not provided
Annual Budget: not provided
Cost to Join: $15

The organization describes itself as follows:

The Animal Legal Defense Fund is a nationwide network of over 250 attorneys dedicated to protecting and promoting animal rights. We are the only advocacy group with the experience and expertise to deliver the legal services that can be crucial to saving animal lives.

ANIMAL LIBERATION NETWORK
P.O. Box 983
Hunt Valley, MD 21030
Telephone: (301) 666-9113

Number of Members: not provided
Year Founded: 1988
Annual Budget: not provided
Cost to Join: $15

The organization describes its goals as follows:

The ALN is a non-profit organization addressing animal rights and related environmental issues. ALN is a total abolitionist organization working to reach the mass general public through a variety of programs and events. It is ALN's goal to attain the complete abolition of all forms of non-human animal abuse and exploitation within our lifetime.

ANIMAL PROTECTION INSTITUTE
2831 Fruitridge Road
P.O. Box 22505
Sacramento, CA 95822
Telephone: (916) 731-5521
Fax: (916) 731-4467

Number of Members: 180,000
Year Founded: 1968
Annual Budget: $3 million
Cost to Join: $20 year

The organization describes its goals as follows:

> To alleviate or eliminate fear, pain and suffering among all animals everywhere.

Among the group's publications are *Product Testing: A Way Without Animals* (30¢) and *Animals in Research* (35¢).

ANIMAL RIGHTS COALITION

P.O. Box 20315
Bloomington, MN 55420
Telephone: (612) 822-6161

Number of Members: 2,000 mailing list, 300 active
Year Founded: 1981
Annual Budget: not provided
Cost to Join: $15

The organization describes itself as follows:

> Nonprofit tax-exempt organization dedicated to eliminating animal suffering and exploitation. Believes that the respect for life is the ultimate moral concern.

ANIMAL RIGHTS INTERNATIONAL

Box 214, Planetarium Station
New York, NY 10024
Telephone: (212) 873-3674

Number of Members: ARI is an umbrella organization for the Coalition to Abolish the LD50, the Coalition to Abolish the Draize and the Coalition for Nonviolent Food.
Year Founded: 1975
Annual Budget: $75,000
Cost to Join: no charge ("Individuals and organizations are encourged to participate in the coalitions at whatever level they feel comfortable.")

The organization describes its goals as follows:

> To phase down and phase out the use of animals in testing, education and research without compromising human safety. To phase down the pain and suffering of animals raised for food so long as people continue to eat them. To promote nonviolent food.

ARI is coordinated by Henry Spira, who produces an annual Coordinator's Report that is available for 65¢ in stamps to cover postage.

ANIMAL WELFARE INSTITUTE

P.O. Box 3650
Washington, DC 20007
Telephone: (202) 337-2332

Number of Members: not provided
Year Founded: 1951
Annual Budget: $440,000
Cost to Join: $15

The organization describes itself as follows:

Interested in the promotion of humane treatment of animals, and incorporated as a non profit organization under the laws of Delaware. Its interests and membership are nationwide. Its purpose is to promote the welfare of all animals and to reduce the total of pain and fear inflicted on animals by man.

The institute says one of its major aims is "humane treatment of laboratory animals and the development and use of non-animal testing methods wherever possible."

AWI's lobbying arm is the Society for Animal Protective Legislation. (See entry below.)

THE ANIMALS' AGENDA MAGAZINE

456 Monroe Turnpike
Monroe, CT 06468
Telephone: (203) 452-0446
Fax: (203) 452-9543

Number of Members: 30,000 circulation
Year Founded: 1979
Annual Budget: $500,000
Cost: $22 a year

The organization describes itself as follows:

The *Animals' Agenda*, published by the Animal Rights Network Inc., is a magazine published 10 times per year, covering a broad range of topics concerning animals, animal rights and the environment. It serves as an outreach vehicle to the general public, a source of vital information to activists, and a forum for rational dialogue.

ARGUS ARCHIVES INC.

228 East 49th Street
New York, NY 10010
Telephone: (212) 355-6140

Number of Members: no members
Year Founded: 1969
Annual Budget: not provided
Cost to Join: no members

The organization describes itself as follows:

ARGUS is an archives/library that deals with the major issues of animal abuse and welfare. Contains approximately 800 folders on various organizations and corporations involved in the issue of animal abuse. Goal is to be available to writers, researchers, and all others who are looking for the facts on the above issues. The center is open to the public by appointment only.

ASSOCIATION OF VETERINARIANS FOR ANIMAL RIGHTS (AVAR)

15 Dutch Street
Suite 500-A
New York, NY 10038-3779
Telephone: (212) 962-7055
Fax: (212) 962-7056

Number of Members: 1,500
Year Founded: 1981
Annual Budget: not provided

Cost to Join: $30 veterinarian
$10 veterinary medical student
$20 veterinary technician or supporting contributor

The organization describes itself as follows:

> AVAR believes that all animals have value and interests independent of the value and interests of other animals, including people. Veterinarians who belong to AVAR actively protect the rights of their patients by providing expert testimony at court trials and legislative hearings, writing letters and speaking on issues that affect non-human animals.

BEAUTY WITHOUT CRUELTY USA
175 West 12th Street, #16G
New York, NY 10011-8275
Telephone: (212) 989-8073

Number of Members: about 7,000
Year Founded: 1972
Annual Budget: $25,000 to $100,000
Cost to Join: $15 regular
$10 senior or student

The organization describes itself as follows:

> Our aims are: 1) To inform the public about the massive suffering of many kinds of animals in the fashion, cosmetics and household products industries. 2) To provide information about which fashions, cosmetics and household products do not involve suffering, confinement or death of any animal in their development or testing. This includes detailed, updated lists of cruelty-free cosmetics, household products and fashions, with outlets where these can be obtained. [BWC USA maintains] a database of stores where [these products] are available.

The group publishes "The Compassionate Shopper," "Action Alert" and "Please Excuse Me for Approaching You—I See You Are Wearing Fur" (all 15¢ each).

CAMBRIDGE COMMITTEE FOR RESPONSIBLE RESEARCH
P.O. Box 1626
Cambridge, MA 02138
Telephone: (617) 547-9255

Number of Members: not provided
Year Founded: 1986
Annual Budget: not provided
Cost to Join: $15 or donation

The organization describes itself as follows:

> CCRR promotes accountability in animal research through public outreach, education, lobbying and investigation of animal cruelty.

In September 1987, the committee issued a study that it says documents "a pattern of neglect and abuse of animals in Cambridge laboratories."

CITIZENS TO END ANIMAL SUFFERING AND EXPLOITATION (CEASE)
P.O. Box 27
Cambridge, MA 02238
Telephone: (617) 628-9030

Number of Members: 3,000
Year Founded: 1978
Annual Budget: $75,000
Cost to Join: $15

The organization describes itself as follows:

> CEASE is an educational and activist organization opposed to all forms of animal abuse and exploitation. We have a multi-media anti-fur campaign using advertising and protests; a Compassion Campaign in which we work with health food stores to promote cruelty-free products; and an education program which includes giving presentations at schools. In 1988 we launched the farm animal referendum which reached the November ballot in Massachusetts.

CIVIS/CIVITAS
P.O. Box 26
Swain, NY 14884
Telephone: (607) 545-6213

Number of Members: "informal"
Year Founded: 1983
Annual Budget: "informal"
Cost to Join: $15 a year

The organization describes itself as follows:

Our purpose is to show the public why vivisection is not a good way to promote human health and to place information in the hands of other activists so that they can make use of it too.

COMPASSION FOR ANIMALS FOUNDATION
P.O. Box 5312
Beverly Hills, CA 90209-5312
Telephone: (213) 204-6600
Fax: (213) 559-8595

Number of Members: not applicable
Year Founded: 1986
Annual Budget: not provided
Cost to Join: not applicable

The organization describes itself as follows:

> Compassion for Animals Foundation seeks to educate the public about the exploitation and abuse of animals in all facets of society through the publication of *The Animals' Voice* magazine, a comprehensive journal that exposes these abuses and promotes a humane, compassionate lifestyle.

The Animals' Voice has a circulation of 20,000. It costs $18 for six issues a year.

DORIS DAY ANIMAL LEAGUE
Suite 200
111 Massachusetts Avenue, N.W.
Washington, DC 20001
Telephone: (202) 842-3325

Number of Members: 300,000

Year Founded: 1987
Annual Budget: $3.9 million
Cost to Join: $10

The organization describes itself as follows:

> The Doris Day Animal League is a nonprofit, nationwide citizens' lobbying organization, formed to focus public attention on the needless suffering of many animals in commercial testing facilities and laboratories. The group provides people with the names of their congressional representative and senators and summaries of the issues. Members and non-members of the Animal League are encouraged to file petitions expressing their concerns with their elected officials—much in the way Americans have spoken out over the years about issues related to the environment, consumer protection, and women's and civil rights.

FEMINISTS FOR ANIMAL RIGHTS
P.O. Box 10017
North Berkeley Station
Berkeley, CA 94709
Telephone: (415) 547-7251

Number of Members: mailing list of 800
Year Founded: 1981
Annual Budget: negligible
Cost to Join: donations

The organization describes itself as follows:

FAR is a group of vegetarian women who are dedicated to ending all forms of animal abuse. We seek to educate people through the dissemination of literature and speaking engagements on the connections between abuse of animals and abuse of women.

FOCUS ON ANIMALS
P.O. Box 150
Trumbull, CT 06611
Telephone: (203) 377-1116

Number of Members: not applicable
Year Founded: 1987
Annual Budget: $35,000
Cost to Join: not applicable

The organization describes itself as follows:

> Produces and distributes video documentaries on a range of animal rights/welfare issues for use in groups, schools and on cable television.

FRIENDS OF ANIMALS
P.O. Box 1244
South Norwalk, CT 06856
Telephone: (203) 866-5223
Fax: (203) 853-9102

Number of Members: 104,000
Year Founded: 1957
Annual Budget: $3.4 million
Cost to Join: $20 year

The organization describes itself as follows:

Friends of Animals is an international, not-for-profit organization that works to reduce and eliminate animal suffering. Today, we are one of the nation's most respected and active animal protection organizations with offices in Connecticut, New York City, Rhode Island, Washington, D.C., Florida, California, Paris and France. FoA's goal is to achieve a humane ethic in the treatment of all animals on this earth. To realize this goal, we identify and investigate inhumane practices, vigorously expose them to public view, and implement programs of action to bring them to an end.

...FoA works to stop the needless and cruel uses of animals for experiments, research and testing wherever they exist. FoA has exposed the frivolity, invalidity and unnecessary repetition in research involving the use of animals.

HUMANE EDUCATION COMMITTEE
P.O. Box 445
Gracie Station
New York, NY 10028
Telephone: (212) 410-3095

Number of Members: 1,500+
Year Founded: 1982
Annual Budget: $15,000
Cost to Join: $15

The organization describes itself as follows:

Promotes institutionalizing hu-mane education programming in school systems grades pre-kindergarten through twelfth.

HUMANE SOCIETY OF THE UNITED STATES
2100 L Street, N.W.
Washington, DC 20037
Telephone: (202) 452-1100
Fax: (202) 778-6132

Number of Members: 400,000
Year Founded: 1954
Annual Budget: $10 million
Cost to Join: $10 year

The organization describes itself as follows:

The primary and motivating concern of The Humane Society of the United States is the prevention of cruelty to all living creatures. We are mindful that man has been uniquely endowed with a sense of moral values. For this reason, we believe he is responsible for the welfare of those animals that he has domesticated and those upon whose natural environment he encroaches. The responsibility, we believe, must be shared by all people. This does not matter if they benefit from the use of such domestic animals or participate in the alteration of environments supporting the life of other creatures. As the dominant intelligent life form on earth, we are accountable as a species. Though we are not opposed to the legitimate and appropriate utilization of animals in the service of man, such utilization gives man neither the right nor the license to exploit or abuse any animal in the process.

Among the organization's principles is the following: "It is wrong to use animals for medical, educational or commercial experimentation or research unless the following criteria are met: absolute necessity; no available alternative methods; and no pain or torment caused to the animals."

IN DEFENSE OF ANIMALS
21 Tamal Vista Boulevard
Corte Madera, CA 94925
Telephone: (415) 924-4454
Fax: (415) 927-2607

Number of Members: 45,000
Year Founded: 1984
Annual Budget: $644,000
Cost to Join: donations

The organization describes itself as follows:

Organization established by veterinarian Elliot Catz. What prompted his desire was a professor of psychology who conducted experiments on animals who eventually died from infection and disease. In Defense of Animals coordinated World Laboratory Animal Liberation Week, which gathers organizations and individuals to protest against animal abuse. Its main focuses have been animal research, pound seizure, their animal abuse hotline, outreach, their anti-fur campaign. Also assist callers

needing help with animals who have injured due to car accidents, abuse, etc.

INTERNATIONAL FOUNDATION FOR ETHICAL RESEARCH
79 West Monroe, Suite 514
Chicago, IL 60603
Telephone: (312) 419-6990

Number of Members: not applicable
Year Founded: 1985
Annual Budget: approx. $150,000
Cost to Join: not applicable

The organization describes itself as follows:

Our mission is to encourage the cooperative efforts of scientists to find alternatives to the use of live animals in research, testing and teaching. We believe the time has come to become more aggressive and imaginative in the search for alternatives that are ethical, valid and cost efficient.

IFER has already sponsored research that is leading the way to replacing the Draize Eye Irritancy Test and the LD50 test....We expect to continue to offer financial assistance to qualified scientists and are pleased to offer graduate fellowships in addition to the annual research grants.

IFER sponsors lectures, workshops, seminars and symposia where ideas will be exchanged among scientists and other interested parties concerning the use of alternative techniques and practices. A quarterly newsletter is produced to announce impending activities and to report on the results of IFER-sponsored programs and other progress in the field.

INTERNATIONAL FUND FOR ANIMAL WELFARE
P.O. Box 193
Yarmouth Port, MA 02675
Telephone: (508) 362-4944
Fax: (508) 362-5841

Number of Members: 650,000+
Year Founded: 1969
Annual Budget: not provided
Cost to Join: donations

The organization describes itself as follows:

IFAW's activities include getting one million signatures protesting seal hunts and presenting them to the Norwegian embassy in the U.S., assisting in the purchase of 10,000 leg-hold traps in the U.S. then destroying them, saving aged huskies from being hung, the purchase of land in Costa Rica to provide a home for bees, stepped up efforts to reduce the importation of white coal seal products by the EEC, and rescuing whales stranded and beached in Cape Cod.

Specifically on animal testing, IFAW recently organized a campaign to get people to send a protest form to cosmetics manufacturers involved with animal testing. The group is also lobbying the Cosmetic,

Toiletry and Fragrance Association to develop a "cruelty-free product logo" for products produced without live animal testing.

INTERNATIONAL NETWORK FOR RELIGION AND ANIMALS
2913 Woodstock Avenue
Silver Spring, MD 20910
Telephone: (301) 565-9132

Number of Members: not provided
Year Founded: not provided
Annual Budget: not provided
Cost to Join: $10 student/senior citizen
                $15 individual
                $25 family
                $500+ patron

The organization describes its concerns as follows:

Far from showing mercy, humanity uses its dominion over other animal species to pen them in cruel close confinement, to trap, club, and harpoon them, to poison, mutilate and shock them in the name of science, to kill them by the billions, and even slowly to blind them in excruciating pain to test cosmetics.

Among its goals are "To encourage and facilitate meetings betwen leaders in the animal welfare and rights movement and leaders in religion for dialogue, periodically, on animal issues."

The organization plans to publish materials for use in educational programs that will be promoted "within all levels of organization of religious groups."

INTERNATIONAL SOCIETY FOR ANIMAL RIGHTS
421 South State Street
Clarks Summit, PA 18411
Telephone: (717) 586-2200

Number of Members: 55,000
Year Founded: 1959
Annual Budget: $600,000
Cost to Join: $15 member
              $50 voting member

The organization describes itself as follows:

> International Society for Animal Rights Inc. (ISAR), the nation's first Animal Rights organization, is opposed to the exploitation of animals and publishes documentary information on the exploitation and suffering of animals in order to enlist public opposition. ISAR, for example, in 1978 launched the campaign against product testing on animals. ISAR led the campaigns against the use of pound/shelter animals for experimentation, achieving repeal of New York State, Connecticut and Los Angeles laws/ordinances which mandated the use of pound/shelter animals. ISAR also drafted and successfully lobbied through New Jersey's law forbidding use of pound/shelter animals for experimentation.

> ISAR is especially concerned about the suffering of animals in laboratories but also exposes and opposes other major causes of animal suffering such as the production and wearing of fur; the use of animals in entertainment (rodeos, circuses, dog and horse racing); hunting, trapping, fishing; the overpopulation of dogs and cats resulting from lack of breeding control and is an advocate of spay/neuter clinics.

> ISAR publishes books, monographs, brochures and the ISAR Report, a 10 times a year newsletter.

LAST CHANCE FOR ANIMALS
18653 Ventura Boulevard
Suite 356
Tarzana, CA 91356
Telephone: (818) 760-2075

Number of Members: not provided
Year Founded: not provided
Annual Budget: not provided
Cost to Join: donations; $25 suggested

The organization describes itself as follows:

> Last Chance for Animals takes direct, positive action to end the hideous abuses of vivisection. We're a total abolitionist antivivisection organization—and in our relatively short life, we've mounted a number of protests and as a consequence, accepted and served jail terms to dramatize our message.

MASSACHUSETTS SOCIETY FOR THE PREVENTION OF CRUELTY TO ANIMALS

350 South Huntington Avenue
Boston, MA 02130
Telephone: (617) 522-7400
Fax: (617) 522-4885

Number of Members: 45,000
Year Founded: 1868
Annual Budget: $15.9 million
Cost to Join: $30

The organization describes itself as follows:

> MSPCA provides the widest range of animal services including three veterinarian hospitals, eight animal shelters, a law enforcement facility and legislative programs affiliated with the American Humane Educational Foundation, which operates a charitable hospital in Morocco and a London-based society for the protection of animals. Relies exclusively on contributions, membership dues, service fees and an endowment.

The MSPCA publishes the bimonthly *Animals Magazine*, available for $15 a year.

MEDICAL RESEARCH MODERNIZATION COMMITTEE

P.O. Box 6036
Grand Central Station
New York, NY 10163-6018
Telephone: (212) 876-1368

Number of Members: 1,200
Year Founded: 1978

Annual Budget: not provided
Cost to Join: $25
$10 students/senior citizens

The organization describes itself as follows:

> The MRMC is a national group, predominantly of health care professionals, who lend their experience and expertise to identify and promote modern methods of research. The MRMC encourages wider use of modern research technologies, such as cat scans, pet scans, needle biopsies, tissue cultures, computer models, and post-market surveillance of drugs, which permit safe, direct study of human disease with human patients.

NATIONAL ALLIANCE FOR ANIMAL LEGISLATION

P.O. Box 75116
Washington, DC 20013
Telephone: (703) 684-0654
Fax: (703) 684-4526

Number of Members: 12,000
Year Founded: 1985
Annual Budget: $200,000
Cost to Join: $25

The organization describes its purpose as follows:

> To lobby for animal protection legislation; to organize and maintain a national network of individuals interested in lobbying on behalf of animals; to provide updated legislative information to

our members.

NATIONAL ANTI-VIVISECTION SOCIETY

53 W. Jackson Boulevard
Suite 1550
Chicago, IL 60604
Telephone: (312) 427-6065

Number of Members: 45,000
Year Founded: 1929
Annual Budget: $1 million
Cost to Join: $10 individual
                  $15 family
                  $5 student
                  $50 life
                  $100 life benefactor

The organization describes itself as follows:

> Responds daily to requests for help and information from members and others around the country who are individually fighting vivisection. In addition, through a national public awareness campaign, NAVS reaches hundreds of thousands of persons each year with its brochures and informational pieces. Also continues to provide assistance to students at every level concerned with instruction, research and tests performed on live animals.

NATIONAL ASSOCIATION FOR HUMANE ENVIRONMENTAL EDUCATION

67 Salem Road
East Haddam, CT 06423
Telephone: (203) 434-8666

Number of Members: not provided
Year Founded: 1973
Annual Budget: not provided
Cost to Join: $20

The organization describes itself as follows:

> The National Association for Humane Environmental Education is the elementary and secondary education division of The Humane Society of the United States. NAHEE serves as a national resource for teachers and other educators concerned about quality education and the teaching of respect and responsibility toward all living creatures and the environment. NAHEE is a nonprofit organization supported through individual contributions, private grants and allocations from the Humane Society of the United States.

NATIONAL ASSOCIATION OF NURSES AGAINST VIVISECTION (NANAV)

P.O. Box 42110
Washington, DC 20015-0110
Telephone: (301) 770-8968

Number of Members: 500
Year Founded: 1985
Annual Budget: not provided
Cost to Join: $15/$10 student

The organization describes itself as follows:

> The NANAV unites members of the nursing profession who are

committed to the abolition of all forms of animal experimentation, and who support a more effective distribution of health care dollars.

As health care professionals and persons committed to the care of others, nurses are in a unique position to voice opposition to vivisection and to support programs that help people rather than harm animals.

NANAV's goals include promoting a more effective allocation of limited health resources by emphazing wellness and the prevention of illness, including promotion of a vegetarian diet, promoting the use of sophisticated and sensible non-animal research methods, informing the public and members of the nursing and medical professions about the enormous suffering of millions of animals used in research, and challenging the myth that the use of animals in medical research is "necessary."

NEW ENGLAND ANTI-VIVISECTION SOCIETY

333 Washington Street
Suite 850
Boston, MA 02108-5100
Telephone: (617) 523-6020
Fax: (617) 523-7925

Number of Members: 10,000
Year Founded: 1895
Annual Budget: $1.3 million
Cost to Join: $10

The organization describes itself as follows:

NEAVS is dedicated to ending the abuse of animals in laboratories. Campaigns focus on public education, legislative change and litigation. We are anti-vivisection, not anti-research. Our goal is to eventually end the use of live animals for research, experimentation, or testing in the belief that it is morally wrong and in most cases, scientifically unsound, to subject animals to vivisection. Scientific and medical knowledge will advance whether or not animals are used in research....Alternative methods of developing and testing products have proven to be more reliable, more efficient and less expensive thanks to efforts by NEAVS [and] similar organizations....NEAVS has funded many activities aimed at re-educating scientists about research alternatives to live animal experimentation. Among them: research grants and informative seminars at major medical schools exploring cell culture and other technologies which promise to offer substitutes for the use of live animals in experimentation. NEAVS is working with physician, veterinary and psychological associations which promote the use of such alternative methods and oppose animal experimentation as outdated.

NORTH CAROLINA NETWORK FOR ANIMALS

P.O. Box 33565
Raleigh, NC 27636
Telephone: (919) 787-7435

Number of Members: 10,000 on mailing list
Year Founded: 1983
Annual Budget: $50,000
Cost to Join: $15

The organization describes itself as follows:

> We are all-volunteer, nonprofit, nonviolent, but firmly committed to animal rights. Our central office is in Raleigh, but we have 10 chapters in major North Carolina cities. We focus on North Carolina issues, but mailing/membership includes 37 other states. Our major goal is to provide information to the public. We speak at universities, schools, clubs. We hold 30-40 protests a year, set up literature tables in each chapter at least twice a year, have a booth at the state fair (for a week). We challenge research that seems duplicative, promote animal protective legislation, assist humane societies, promote meat-free (veal free) restaurant offerings and school meals and promote spaying and neutering and beaver relocation—working on any animal issue at any local level. We have filed suit against [the Department of Agriculture's Animal and Plant Health Inspection Service] to help stop pet theft. The issues we promote are promoted actively.

The group publishes a newsletter six times a year and is preparing for publication a special general animal issues color brochure.

PEOPLE FOR THE ETHICAL TREATMENT OF ANIMALS (PETA)

P.O. Box 42516
Washington, DC 20015
Telephone: (301) 770-7444
Fax: (301) 770-8969

Number of Members: 250,000
Year Founded: 1980
Annual Budget: $5.2 million
Cost to Join: $20

The organization describes itself as follows:

> Educational and activist group opposed to all forms of animal oppression and exploitation. PETA combats speciesism by holding weekly activist workshops, giving lectures and showing films,...picketing, street theatre, investigations (undercover).

> PETA has been the coordinator of the shareholder campaign to get consumer products companies to report on their involvement with animal testing.

> PETA runs a student group called Students for the Ethical Treatment of Animals.

PHYSICIANS COMMITTEE FOR RESPONSIBLE MEDICINE
P.O. Box 6322
Washington, DC 20015
Telephone: (202) 686-2210

Number of Members: 25,000
Year Founded: 1985
Annual Budget: $134,000
Cost to Join: $20

The organization describes itself as follows:

> The expertise of physicians is employed to promote alternatives to the use of animals in research and testing.

PROGRESSIVE ANIMAL WELFARE SOCIETY (PAWS)
P.O. Box 1037
Lynnwood, WA 98946
Telephone: (206) 743-1884

Number of Members: 10,000
Year Founded: 1967
Annual Budget: $600,000
Cost to Join: $10

The organization describes itself as follows:

> Protection of all animals. Education of the public concerning the exploitation of animals. Operate full service animal care center, including wildlife rehabilitation. Active Animal Rights Department.

PSYCHOLOGISTS FOR THE ETHICAL TREATMENT OF ANIMALS
P.O. Box 87
New Gloucester, ME 04260
Telephone: (207) 926-4817

Number of Members: 550
Year Founded: 1981
Annual Budget: $20,000-$40,000
Cost to Join: $25

The organization describes itself as follows:

> PsyETA is an independent association of psychologists dedicated to the promotion of animal welfare within the science and profession of psychology.

> Our primary objective is to lessen the suffering of animals in research laboratories, educational and therapeutic settings.

> PsyETA stands for an ethical dimension, balancing the value of experimentation against the suffering of animals. While recognizing the benefits of research, we hold that the rights and interests of the non-human animals involved are substantial and must be respected.

SCIENTISTS CENTER FOR ANIMAL WELFARE (SCAW)
4805 St. Elmo Avenue
Bethesda, MD 20814
Telephone: (301) 654-6390

Number of Members: 1,000

Year Founded: 1978
Annual Budget: $250,000
Cost to Join: $35 individual member
$500-$1,000 institutional
member

The organization describes itself as
follows:

SCAW is a unique, non profit organization of scientists and others
concerned with the well-being of
research animals. Educational
materials will be available to promote responsible and caring
treatment of laboratory, farm and
wildlife research animals; to comply with federal mandates; to
encourage consideration of alternatives.

SOCIETY FOR ANIMAL PROTEC-
TIVE LEGISLATION

P.O. Box 3719
Washington, DC 20007
Telephone: (202) 337-2332
Fax: (202) 338-9478

Number of Members: 13,000 mailing
list
Year Founded: 1955
Annual Budget: $123,527
Cost to Join: donations

The organization describes itself as
follows:

Goal is to work on legislation in
connection with animal welfare.

This group is the lobbying arm of
the Animal Welfare Institute. (See
entry above.)

TRANS-SPECIES UNLIMITED

P.O. Box 1553
Williamsport, PA 17703
Telephone: (717) 322-3252

Number of Members: 30,000
Year Founded: 1981
Annual Budget: $500,000
Cost to Join: $15

The organization describes itself as
follows:

A national grass-roots animal
rights organization dedicated to
building a social movement
against animal exploitation and
focusing on large-scale institu-
tionalized abuse and slaughter of
animals.

Trans-Species Unlimited has been
most visible in organizing the
demonstrations that led Cornell
Medical College to end barbitu-
rate addiction experiments on
cats.

UNITED ACTION FOR ANIMALS

205 East 42nd Street
New York, NY 10017
Telephone: (916) 429-2457
Fax: (916) 867-7283

Number of Members: 20,000+
Year Founded: 1962
Annual Budget: not provided
Cost to Join: donations

The organization describes itself as
follows:

Founded basically to deal with

the alternatives to animal research in the medical arena. Published "The Alternatives to Animal Testing," and other information showing the inhumane experiments, the cost to conduct them, and the fact that they are unnecessary. Feels that most of the experiments are "fiddling around," and serve no specific purpose. UAA feels that the $7 billion the National Institute of Health lends for animal research could be used for the social issues concerning animals.

UNITED ANIMAL NATIONS
5892 A South Land Park Drive
Sacramento, CA 95822
Telephone: (916) 429-2457
Fax: (916) 429-2456

Number of Members: 6,000
Year Founded: 1987
Annual Budget: $700,000
Cost to Join: $25

The organization describes its major goals as follows:

> To unify the efforts of the humane movement, fund worthy grassroots projects, provide emergency disaster assistance to animals and educate about animal protection.

The group publishes the *Journal of United Animal Nations*, which is free with membership in the organization.

Appendix B:
Trade Associations

THE COSMETIC, TOILETRY AND FRAGRANCE ASSOCIATION
1110 Vermont Avenue N.W.
Suite 800
Washington, DC 20005
Telephone: (202) 331-1770

The organization describes itself as follows:

The national trade association representing the cosmetic, toiletry and fragrance industry. CTFA, founded in 1894, has an active membership of approximately 250 companies that manufacture or distribute the vast majority of the finished cosmetic products marketed in the United States. In addition, CTFA includes more than 240 associate member companies from related industries, such as manufacturers of cosmetic raw materials and packaging materials.

CTFA released a position statement on animal testing on Sept. 15, 1989. In part, the statement reads:

The manufacturers of personal care products regard the safety of their consumers as their primary responsibility and highest priority....
Animal testing must remain an option for new ingredients and for new untested combinations of existing ingredients. Each company makes its own scientific judgments concerning the extent, if any, of animal testing it conducts or may require....
The industry...strongly opposes legislation that would remove the option of animal-based research in ensuring human health and safety....
The personal care products industry has been actively supporting research into alternatives to animal testing for the past 10

years, and is very pleased that significant advances have been made in this area. Given the nature of nonanimal screening mechanisms, however, some may apply to one product category while not being suitable for another.

Our members are committed to continuing the search for alternatives, to using the fewest animals possible consistent with human health and safety, and to communicating openly and fully in response to the public's interest in this important subject.

CHEMICAL MANUFACTURERS ASSOCIATION
2501 M Street, N.W.
Washington, DC 20037
Telephone: (202) 887-1100

NATIONAL ASSOCIATION FOR BIOMEDICAL RESEARCH
Foundation for Biomedical Research
818 Connecticut Avenue N.W.
Suite 303
Washington, DC 20006
Telephone: (202) 857-0540

The NABR is the lobbying arm of this organization; it describes its mission as follows:

To activate and represent member institutions in national policymaking which affects the use of animals in research, education, and product safety testing.

The FBR, the tax-exempt part of this group, describes itself as follows:

To assist the public in understanding the importance of humans and responsible animal research in the development of treatments and cures for diseases, disorders and injuries which affect humans and animals, to counter the claims of the animal rights groups which seek to eliminate this research.

A representative of the NABR told IRRC that the group has 320 institutional members; she said 70 percent are mostly universities with medical schools or medical research programs and 30 percent are mostly pharmaceutical companies.

THE SOAP AND DETERGENT ASSOCIATION
475 Park Avenue, South
New York, NY 10016
Telephone: (212) 725-1262

The organization describes itself as follows:

The Soap and Detergent Association is the trade association of the detergent industry. The 148 company members include manufacturers of soaps and detergents and the ingredients for them. Association members are responsible for over 90 percent of U.S. detergent production.

On the issue of animal testing, the SDA told IRRC:

The association and its members have reduced reliance on animal testing and are doing only the testing that is necessary for the safety evaluation of new products. Efforts to reduce reliance on animal testing are continuing. Complete elimination of animal testing for the safety evaluation of new cleaning products does not appear to be possible in the near future.

The association is in the third phase of evaluating a number of candidate nonanimal tests for eye irritancy. These tests have potential for being members of a battery of tests for eye irritancy and for reducing dependency on the Draize test. Other, different tests will most likely need to be identified and evaluated before a complete battery of tests for eye irritancy can be assembled.

Appendix C:
Alternatives Research Centers

JOHNS HOPKINS CENTER FOR ALTERNATIVES TO ANIMAL TESTING (CAAT)
615 North Wolfe Street
Baltimore, MD 21205
Telephone: (301) 955-3343

Dr. Alan Goldberg, director of CAAT, describes his organization as follows:

CAAT was established in 1981 for the purpose of developing and disseminating basic scientific knowledge concerning innovative non-whole animal methods to evaluate fully the safety of commercial and therapeutic products. It is the sole organization of this kind in the United States. The center's mission is to (1) encourage research that will provide knowledge leading to in vitro test procedures or other nonanimal test procedures to examine the toxicity of xenobiotics and commercial products; (2) to develop specific methodologies that will provide alternative approaches to whole animal studies for the evaluation of safety; (3) to disseminate research progress through symposia, publications and workshops; and (4) to encourage technology transfer and the use of in vitro methods in regulatory consideration; and (5) product development.

CAAT publications include a newsletter, published three times a year, which is free upon request. CAAT also published a book series, *Alternative Methods in Toxicology*.

CENTER FOR ANIMALS AND PUBLIC POLICY
Tufts School of Veterinary Medicine
200 Westboro Road
N. Grafton, MA 01536
Telephone: (508) 839-5302

Dr. Andrew Rowan of the center is

now coordinating The Alternatives Project, a two-pronged effort that involves publishing a newsletter, *The Alternatives Report*, which is "intended to provide news and analysis on the latest developments concerning the search for alternatives, on relevant policy issues, and on the role of government, industry and the public. The intended audience includes regulators, industrial toxicologists, other laboratory scientists, and individuals with a serious interest in the alternatives field." The newsletter is currently free of charge.

The other part of the center's project "involves research for a consensus national agenda to support and implement the development and validation of alternatives for safety testing. This may be an idealistic goal but we are pursuing a process that will, at least, focus debate more closely on the alternatives issue." The center is holding a series of workshops and meetings with representatives from industry, the activist community and government regulators. Following these meetings, the center "will produce written drafts of the discussions...and make them widely available for comment and discussion."

THE FUND FOR THE REPLACEMENT OF ANIMALS IN MEDICAL RESEARCH (FRAME)
Eastgate House
34 Stoney Street
Nottingham NG1 1NB
United Kingdom

FRAME describes itself as follows:

FRAME was founded in 1969 to seek a significant reduction in the need for live animal experimentation by promoting an objective and scientific reappraisal of the aims and methodologies of biomedical science and the development, validation and adoption of alternative techniques. FRAME is not an antivivisection society and considers that the immediate and total banning of all live animal experimentation is an unrealistic and unachievable goal, since continued progress in medical and veterinary research, diagnosis and treatment is necessary if the remaining diseases which lessen the length and quality of human and animal life are to be overcome. FRAME also recognizes that new chemicals, including drugs, cosmetics and toiletries, household chemicals and agrochemicals, must be tested, so that possible hazard to those involved in their manufacture or use, or who may be accidentally exposed to them, can be recognized and suitable precautions taken....FRAME seeks to avoid confrontation and to work positively with scientists in academic institutions, in industry, and in government departments, including the regulatory authorities.

FRAME publishes *ATLA*, a publication that the Tufts center describes as "containing news items, original articles, reviews, meeting reports, book reviews and a set of recent references that are relevant to the subject covered." Articles are received and printed from inside and outside the United Kingdom. The North American editor of the journal is Dr. Oliver Flint, Bristol-Myers Squibb, Pharmaceutical Research and Development Division, P.O. Box 4755, Syracuse, NY 13221.

SELECTED BIBLIOGRAPHY

Alperson, Myra. "Animal Advocacy: More Bark, More Bite." *Business and Society Review*, summer 1988, pp. 26-30.

Alternative Methods in Toxicology, Vols. 1-6. New York: Mary Ann Liebert, 1983-1988.

The Alternatives Report. Published by the Center for Animals and Public Policy, Tufts University School of Veterinary Medicine, North Grafton, Mass., 1989.

AMA White Paper. *Use of Animals in Biomedical Research: The Challenge and Response.* Chicago: The American Medical Association, 1989.

The Animals' Agenda. Published by Animal Rights Network Inc., Monroe, Conn.

The Animals' Voice. Published by Compassion for Animals Foundation Inc., Culver City, Calif.

"A Pivotal Year for Lab Animal Welfare." *Science*, Vol. 232, No. 4747 (April 11, 1986), pp. 147-150.

Balls, Michael. "The OTA Report: A Critical Assessment," *ATLA* 14, 1987, pp. 289-297.

Balls, Michael; Riddell, Rosemary J.; and Worden, Alastair N. *Animals and Alternatives in Toxicity Testing.* London: Academic Press, 1983.

Basu, Janet Else. "Improving the Lot of the Laboratory Animal." *The Scientist*, Jan. 9, 1989, pp. 1-3, 20.

Begley, Sharon. "Liberation in the Labs." *Newsweek*, Aug. 27, 1984, pp. 66-67.

Benford, Diane J. "Biological Models as Alternatives to Animal Experimentation." *ATLA* 14 (1987), pp. 318-328.

Bennett, George. "PETA's True Believers and Their Dream." *Montgomery Journal*, Nov. 20, 1989. p. 1, 4-5.

"Beyond the Draize Test." Washington, D.C.: Physicians Committee for Responsible Medicine, 1988.

Bishop, Katherine. "From Shop to Lab to Farm, Animal Rights Battle Is Felt." *The New York Times*, Jan. 13, 1989, pp. A1, A7.

Bishop, Katherine. "Growing Militancy for Animal Rights Is Seen." *The New York Times*, Jan. 19, 1988.

Blumenstyk, Goldie. "With State Legislatures as the Battleground, Scientists and College Officials Fight Animal Welfare Groups." *The Chronicle of Higher Education*, Vol. 35, No. 30 (April 5, 1989), pp. 1, 26.

Bohlen, Celestine. "Animal Rights Case: Terror or Entrapment?" *The New York Times*, March 3, 1989, p. B1, B4.

Buffa, Denise. "2nd Informant Revealed in U.S. Surgical Bomb Plot." *The Advocate* (Stamford, Conn.), Jan. 27, 1989.

Burch, R.L. and Russell, W.M.S. *The Principles of Humane Experimental Technique.* London: Methuen, 1959.

"Cambridge Adopts Animal Research Rules." *The Washington Post,* June 28, 1989.

Capsis, John. "Feds Investigate Trutt Sting." *Westport News* (Westport, Conn.), Feb. 1, 1989.

Capsis, John. "Trutt's 'Friend' Tied to Perceptions." *Westport News* (Westport, Conn.), Jan. 27, 1989.

Carey, John. "Will Relief for Lab Animals Spell Pain for Consumers?" *Business Week*, Oct. 30, 1989, pp. 43-44.

"Consumer Product Safety Commission Animal Testing Policy." *Federal Register*, Vol. 49, No. 105 (May 30, 1984), p. 22522.

Cowley, Geoffrey. "The Battle Over Animal Rights: A Question of Suffering and Science." *Newsweek*, Dec. 26, 1989, pp. 50-59.

Dalton, Rex. "Waging War On The Animal Rights Lobby." *The Scientist*, Feb. 6, 1989, pp. 1, 4.

Feder, Barnaby J. "Beyond White Rats and Rabbits." *The New York Times*, Feb. 28, 1988.

Feder, Barnaby J. "Pressuring Perdue." *The New York Times Magazine*, Nov. 26, 1989., p. 32-34, 60, 72.

Feder, Barnaby J. "Research Looks Away From Laboratory Animals." *The New York Times*, Jan. 29, 1989, p. E24.

Federal Hazardous Substances Act Regulations. 16 CFR, Ch. II, Part 1500 (Jan. 1, 1989), pp. 333-343.

"Food and Drug Administration LD50 Test Policy." *Federal Register*, Vol. 53, No. 196 (Oct. 11, 1988), p. 39650.

Fox, Michael Allen. *The Case for Animal Experimentation: An Evolutionary and Ethical Perspective*. Berkeley: University of California Press, 1986.

Frank, Peter H. "Faint Praise for Noxell's Testing Policy." *The Baltimore Sun*, Jan. 22, 1989, Sec. D, p. 1-2.

Frazier, John M. and Goldberg, Alan M. "Alternatives to Animals in Toxicity Testing." *Scientific American*, Vol. 261, No. 2 (August 1989), pp. 24-30.

Gallagher, Patricia. "Firms Face a Spreading Animal Rights Fight." *The Cincinnati Enquirer*, Oct. 23, 1988, Sec. I, pp. 1,3.

Gallagher, Patricia. "P&G Proposes Animal Test Group." *Cincinnati Enquirer*, July 26, 1989, p. B7.

Gettings, Stephen D. "CTFA Evaluation of Alternatives Program." Unpublished paper, Cosmetic, Toiletry and Fragrance Association, Washington, D.C., 1989.

Greenfield, Meg. "In Defense of the Animals." *The Washington Post*, April 11, 1989.

Goldberg, Alan M. "Animals and Alternatives in Toxicology." Unpublished paper, Johns Hopkins University School of Hygiene and Public Health, Baltimore, Md., 1989.

Great Britain. Home Office. *Statistics of Scientific Procedures on Living Animals*. London: Her Majesty's Stationery Office, 1988.

Havemann, Judith. "Animal Rules Uncage Scientists' Complaints." *The Washington Post*, July 9, 1989, pp. A1, A9.

Haywood, Susan. "Attitudes to Animal Usage and the Review System in the U.S." *ATLA* 14 (1987), pp. 369-374.

Henderson, Keith. "Household Products and Laboratory Animal Testing." *The Christian Science Monitor*, March 20, 1986, p. 33.

Henderson, Keith. "Reducing Animal Testing." *The Christian Science Monitor*, Feb. 4, 1986.

Hoch, David. "Business Ethics, Law, and the Corporate Use of Laboratory Animals," *Akron Law Review*, Vol. 21, No. 2 (Fall 1988), pp. 201-244.

Howe, Marvine. "Advocate for Animal Rights Pleads Guilty in Bomb Case." *The New York Times*, July 15, 1989.

Indiana, Gary. "All Things Cruel and Profitable." *The Village Voice*, Dec. 13, 1988.

"Industry Workshop—Summary Report." Unpublished paper, Tufts Center for Animals and Public Policy, Tufts University School of Veterinary Medicine, North Grafton, Mass., Oct. 25, 1989.

Jackson, Edward M., compiler. *International Directory of Contract Laboratories*. 2d ed. New York: Marcel Dekker, Inc., 1989.

Johns Hopkins Center for Alternatives to Animal Testing newsletter. Published by CAAT, Baltimore, Md.

Keenan, Marney Rich. "Of Mice and Men." *The Detroit News*, May 2, 1989.

"Lawyer Would Shield Bomb Suspect from Media." *Telegram* (Bridgeport, Conn.), Feb. 2, 1989.

LaRussa, Robert. "Firms Try Nonanimal Tests." *Women's Wear Daily*, Feb. 24, 1989.

Loeb, Jerod M.; Hendee, William R.; Smith, Steven J. and Schwarz, M. Roy. "Human vs. Animal Rights: In Defense of Animal Research." *Journal of the American Medical Association*, Vol. 262, No. 19 (Nov. 17, 1989), pp. 2616-2720.

Lyall, Sarah. "Pressed on Animal Rights, Researcher Gives Up Grant." *The New York Times*, Nov. 22, 1988, pp. B1, B5.

Lyall, Sarah. "Animal-Rights Suits Opening Up Research Panels." *The New York Times*, Aug. 22, 1989, pp. B1, B4.

McGill, Douglas C. "Cosmetic Companies Quietly Ending Animal Tests." *The New York Times*, Aug. 2, 1989, p. A1, D22.

Mehlman, M.A., ed. *Benchmarks: Alternative Methods in Toxicology*. Princeton: Princeton Scientific Publishing Co., 1989.

"The Money-Guzzling Genius of Biotechnology." *The Economist*, May 13, 1989, pp. 69-70.

Morgan, Pat. "Animal Instinct." *Detroit Free Press*, Aug. 13, 1989, Sect. J., pp. 1, 4-5.

Nethery, Lauren B. and McArdle, John E. *Animals in Product Development and Safety Testing: A Survey*. Washington, D.C.: The Institute for the Study of Animal Problems, 1985.

Ottoboni, M. Alice. *The Dose Makes the Poison*. Berkeley: Vincente Books, 1984.

Peterson, Iver. "Meeting Offers Hope for Fewer Animal Tests." *The New York Times*, April 12, 1987.

PETA News. Published by People for the Ethical Treatment of Animals, Washington, D.C.

"Proceedings of the Joint Government-Industry Workshop on Progress Towards Non-animal Alternatives to the Draize Test." *Journal of Toxicology: Cutaneous and Ocular Toxicology*, Vol. 8, No. 1 (March 1989).

Ravo, Nick. "U.S. Surgical Admits Spying on Animal-Rights Groups." *The New York Times*, Jan. 26, 1989.

Regan, Tom. *The Case for Animal Rights*. Berkeley: University of California Press, 1983.

Reitman, Judith. "Surgical Procedures." *Fairfield County Advocate* (Conn.), Nov. 28, 1988, pp. 3, 8.

"Report of the FRAME Toxicity Committee," in Balls, et al., *Animals and Alternatives in Toxicity Testing*, 1983, pp. 501-540.

Rowan, Andrew N. *Of Mice, Models, and Men*. Albany: State University of New York Press, 1984.

Rowan, Andrew N. "Scientists Must Help Find Ways to Give the Public a Significant Role in Overseeing Animal Research." *The Chronicle of Higher Education*, Nov. 29, 1989, Sec. 2, pp. 1-2.

Rowan, Andrew N. "Scientists Should Institute and Publicize Programs to Reduce the Use and Abuse of Animals in Research." *The Chronicle of Higher Education*, April 12, 1989, Sec. 2, pp. 1-2.

Sanders, Greg. "Is Animal Testing Really Necessary?" *Moonbeams*, October 1988, pp. 6-8. (*Moonbeams* is an internal publication of Procter & Gamble.)

Scientists Center for Animal Welfare. *Science and Animals: Addressing Contemporary Issues*. Washington, D.C.: 1988.

Shabecoff, Philip. "Industry Fights Use of Animal Tests to Assess Cancer Risk." *The New York Times*, July 25, 1989, p. C4.

Singer, Peter. *Animal Liberation: A New Ethics for Our Treatment of Animals*. New York: Avon Books, 1977.

Singer, Peter, ed. *In Defense of Animals*. New York: Harper & Row, 1986.

Snyder, Marvin. "We Have a Moral Obligation to Continue Conducting Research Involving Animals." *The Chronicle of Higher Education*, Jan. 18, 1989, Sec. 2, p. B1-B2.

Specter, Michael. "Animal-Research Labs Increasingly Are Under Siege." *The Washington Post*, May 30, 1989, p. A1, A6.

Sperling, Susan. *Animal Liberators: Research and Morality*. Berkeley: University of California Press, 1988.

Spiegel, Marjorie. *The Dreaded Comparison: Human and Animal Slavery*. Philadelphia: New Society Publishers, 1988.

Stephens, Martin L. *Alternatives to Current Uses of Animals in Research, Safety Testing, and Education*. Washington, D.C.: The Humane Society of the United States, 1986.

Stevenson, Richard W. "A Campaign For Research On Animals." *The New York Times*, Jan. 20, 1989.

U.S. Congress, House Committee on Energy and Commerce, Subcommittee on Health and the Environment. *Consumer Product Safe Testing Act: Hearing on H.R. 1635*, 100th Congress, 2nd Session, 1988.

U.S. Congress, House Committee on Small Business, Subcommittee on Regulation and Business Opportunities. *Potential Health Hazards of Cosmetic Products*, 100th Congress, 2nd Session, 1988.

U.S. Congressional Office of Technology Assessment. *Alternatives to Animal Use in Research, Testing, and Education*. Washington, D.C.: Government Printing Office, 1986.

U.S. Department of Agriculture, Animal and Plant Health Inspection Service, Regulatory Enforcement and Animal Care. *Animal Welfare Enforcement*. Washington, D.C.: Government Printing Office, 1987, 1988, 1989.

U.S. Department of Agriculture, Animal and Plant Health Inspection Service, Veterinary Services. *Animal Welfare: List of Registered Research Facilities*. Washington, D.C.: Government Printing Office, 1987, 1988, 1989.

U.S. Department of Health and Human Services, Public Health Service, Food and Drug Administration. *Requirements of Laws and Regulations Enforced by the U.S. Food and Drug Administration*. Washington, D.C.: U.S. Government Printing Office, 1989.

U.S. Environmental Protection Agency. "Alternative Methodology for Acute Toxicity Testing." Unpublished paper, Office of Pesticides and Toxic Substances, Sept. 22, 1988.

Use of Laboratory Animals in Biomedical and Behavioral Research. Washington, D.C.: National Academy Press, 1988.

Valentine, Paul W. "U.S. Accused of Trying to Smear Animal Rights Groups." *The Washington Post*, Dec. 7, 1989, p. A17.

Valentine, Paul N. "Three Animal Rights Activists Charged With Felonies." *The Washington Post*, July 26, 1989.

Walker, Alice. "Am I Blue? Thoughts on Animal Feelings, Human Rights and Justice For All." *The Utne Reader*, January/February 1989, pp. 98-102.

Weiss, Rick. "Inventing the Skin You Love to Test." *The Washington Post*, Feb. 21, 1988, p. B3.

Weiss, Rick. "Test Tube Toxicology." *Science News*, Vol. 133 (Jan. 16, 1988), pp. 42-45.

White, George. "Animal Rights Activists Fight Firms Like Cats and Dogs." *The Los Angeles Times*, Dec. 10, 1989.